Imprint

NATIONS, DO NOT WAIT FOR THE ENERGY CORPORATIONS!
We can reduce our electricity price and the expenditures for heating and driving our car by two thirds. How? By consuming our own power coming from our own photovoltaic modules and wind power shares, from "photovoltaic plug-in fields" in our county.
The proof that this works even without feed-in tariffs, can be found in this book!
By Clemens Hauser

Published by: Clemens Hauser Publishing, Balgheim, Germany
Copyright: © 2013 Clemens Hauser, Balgheim, Germany, July 2013
1st edition

Composition:
Gottfried & Simmer Digital Publications GbR, Berlin, ebookatelier.com

Cover design:
Liv Mann

ISBN POD 978-3-9816074-5-1
ISBN Epub 978-3-9816074-1-3
ISBN Mobi 978-3-9816074-3-7

Clemens Hauser

NATIONS, DO NOT WAIT FOR THE ENERGY CORPORATIONS!

We can reduce our electricity price and the expenditures for heating and driving our car by two thirds. How? By consuming our own power coming from our own photovoltaic modules and wind power shares, from "photovoltaic plug-in fields" in our county.
The proof that this works even without feed-in tariffs, can be found in this book!

For all people with a sense of responsibility for society and for the planet.
For our children.

"Let us not forget, that we all have a daily "gushing oil well" in the form of light and wind in our garden. Not to use this wind and light capital, would really be the largest waste of energy of all times. But we know better and we can do better!"
Clemens Hauser

Contents

- 1. Discussion and Allocation of Competences in the Parliament
- 2. Beginning of the Planning in a Parliamentary Committee
  - Choosing of an Expert Commission
  - Providing of the Frame Conditions
- 3. Detailed Planning of the Implementation by the Commission of Experts
  - Planning of the Advertising and of the Continuous Information About the Project
  - Selection of the Field-test Counties
  - Conducting of the Tests
- 4. Discussion of the Results in the Committee and in the Parliamentary Plenum Plus Voting Concerning Open Decisions in the Parliament
- 5. Implementation on a Large Scale
- 6. Constant Follow-up and Adjustments If Necessary

Chapter 4 – We Want to Enforce It

Appendices
The author
Bibliography
Illustrations

Preface

First of all, thank you for being interested in solving major challenges that humanity is facing today and, for this reason, for having decided to read this book!

The following pages include a strategy with concrete measures as to how we can reduce the household price of electricity significantly: to 8-9 euro cents (kWh), or respectively to 10-12 dollar cents (kWh). Furthermore, we will see how this is done using the renewable energy sources photovoltaics plus wind power, and we will learn how this all is possible without any cross subsidizing via feed-in tariffs or similar costly means.

Whether your are a citizen interested in this topic, a local politician or a politician on the national level, this book is for all people with a sense of responsibility for society and for everybody who sees that the time to act is now rather than the day after tomorrow. It is a proposal, a guide, and a tool to achieve several important things at the same time – what could be more efficient than that?

This book delivers the proof that we already dispose of everything necessary for a sunny and bright future with cheap and clean energy, and it will motivate you to support this future.

Take part with your heart and your mind and profit from the, hopefully soon, implementation of this concept.

I wish you a lot of inspiration and joy while reading!

Chapter 1

We Can Produce Our Electricity at Half or at Even a Third of the Current Price (Depending on the Actual Price Level in a given Country), Yet So Far, We Have Not Noticed That This Is Possible.

What Are Our Needs? What Needs to Be Done?

The current household price for electricity in many countries is already high and continues to increase. In the following table, you can see these prices for the EU states in 2012 (If your country is not listed, try and find your country's household electricity price in order to see how high your electricity cost is and thus, to have a comparative figure.):

Country	Euro cents/kWh
Belgium	23.27
Bulgaria	8.46
Czech Rep.	14.97
Denmark	29.97
Germany	25.95
Estonia	10.96
Ireland	21.45
Greece	13.91
Spain	18.22
France	14.12
Italy	21.86
Cyprus	27.81
Latvia	13.89
Lithuania	12.60
Luxembourg	16.96
Hungary	15.83
Malta	17.00
Netherlands	no data
Austria	19.75
Poland	14.18
Portugal	19.93
Romania	10.50
Slovenia	15.42
Slovakia	17.16
Finland	15.49
Sweden	20.27
United Kingdom	16.82
Norway	18.81

(Federal Ministry of Economics and Technology 2013, 30a)

(Throughout the book the "euro" currency is used. In order to convert the values into your currency, please use one of the many currency converters that can be found on the Internet.)

Most probably, the price for electricity in your country has also gone up over the past ten years due to the energy corporations' price increases, higher prices for energy production resources or due to an increase of national taxes on energy.

To offer a solution to this high-price energy problem, the following market concept will show how the citizens, the communities and the state, together, could manage the following, within a maximum of two to five years:

- citizens can reduce their electricity price to 8-9 euro cents/kWh
- electricity will then be the most cost-efficient solution for heating and even more cost efficient for electric cars
- your country could switch to 80-100% clean energy, so that the turn to a sustainable energy supply could soon be finalized
- the new boom of clean energy production is possible even without any feed-in tariffs.

Sounds ambitious? With the right plan and the right planning everything can be accomplished! Don't you think so as well?

So, how do we achieve all that?

We will achieve the above-mentioned objectives through **"photovoltaic plug-in fields"** with **automatic solar tracking,** which are provided **virtually free of charge** by the **county/rural district** for the citizens and into which the **naked photovoltaic modules and batteries** will simply be **plugged in.**

As soon as the modules and batteries are plugged in, they deliver **self-created and self-owned energy** directly to the electric meter at home and at a **long-term low price** that only costs **a fraction of the current household electricity price.**

For the citizens, this means: **Everyone can then produce his own low-price electricity,** regardless of whether he is a **house owner or a tenant.**

The photovoltaic modules and batteries can be ordered directly via the county at the **factory price,** and they can be financed, with the help of the county's staff, **through the national development bank,** like e.g. the KfW bank in Germany. The procurement of the modules and batteries takes place with the **concentrated buying power of all counties in your country** as a result of which a respective discount from the manufacturers is obtained. This discount is passed on to the citizens **without any markup.** The financing occurs at **0.5-1% interest** and with small, **self-determined monthly rates.**

In addition to saving money, there are even **opportunities to earn additional money** with the produced energy. The electricity surplus of the participants can be **interchanged within the county community,** or, **organized by supra-regional cooperatives, it can be sold at the electricity stock exchange.** In order for the private producers to achieve the **most favorable price at the electricity stock exchange,** the electricity will be **intermediately stored – in order to achieve a stable flow –**

by the cooperatives.

For the **colder and darker season** (in case your country has a cold season) a certain amount of the self-produced energy is **stored** in **gas form,** converted via the **Power-to-Gas method.** The facility where it is stored is the **national gas grid.** In this way, even in the dark season, you can profit from very **cheaply generated photovoltaic energy.** The storing is administered by the cooperatives. Additionally, the county sells **shares of wind turbines and erects them according to demand** in order to **relieve** the gas grid and the night-storage batteries again.

Now imagine how, through this concept, we will have clean and cheap electricity and even a surplus of it! Imagine how we consume it, heat with it and drive our e-car with it, as much as we want and with a clear environmental conscience! And, imagine that we can even sell it and thus minimize electricity costs, this means we can have a small additional income by doing so. All this we can achieve because we have an "oil-well" on our land, a sufficient and clean "well" in the form of light/sun and wind, which provides a constant flow of energy for your home, your car and for selling. With the idea described above you then own your own private power plant, with which you could even some day move to another country or merely into another county by just taking it with you (the shares of the wind turbines will then of course be refunded accordingly).

In the following section, you will find out even more reasons why the solution described in this book is realizable, can be implemented quickly and be implemented on a large scale.

(Throughout the book I will often make reference to Germany as it has lively competition between the conventional energy producing industry and renewable energy production. I further make reference to Germany in order to surprise you because, despite the comparatively mediocre sunshine level it has, most probably your country has more sunshine, photovoltaics have become, even there, the most cost-efficient means of energy production. But more about that later.)

So Why Photovoltaic?

The most important key for our project is the fact that the prices of photovoltaic modules have continuously decreased and therefore the kWh of photovoltaic electricity has become a lot cheaper. Another important factor is that we just need the naked modules for our concept. This keeps the electricity generation price low in comparison to roof-top installations, where many more costs add to the electricity price. Using our market model, one kWh of photovoltaic electricity consequently is much cheaper than the household electricity price of the electric supply companies (In comparison: The household electricity price in Germany 2013 is currently at 27.3 cents/kWh [Verivox 2013] – the self-produced photovoltaic energy, calculated on the basis of a product life of the photovoltaic cells of 30 years, ranges at a third of the price and this at the moderate German level of solar radiation.).

We will see in the following example exactly how little your own photovoltaic power will actually cost when it is delivered for your private usage from the plug-in field. For reasons of clarity we set up a very simple calculation.

According to an assessment of the "Fraunhofer Institute for Solar Energy Systems", a combination of standard modules of 1 kW capacity (e.g. 5 standard modules of 200 Watt-peak) is able to produce 1,100 kWh at an average solar irradiance level of 1,300 kWh per square meter and year (Kost et al.

2012, p. 11). This is the solar radiation level that you have for example in southern Germany (to find out about the solar radiation level in your country and the resulting photovoltaic power output, please have a look in the appendix section at the end of this book). If the technique of automatic solar tracking is additionally used, we can increase the output by up to another 30% (Wirth 2013, p. 52), i.e. we achieve 1,430 kWh.

Calculated for 30 years and applying an average capacity loss of the module per year of conservatively put 0.5% (Wirth 2013, p. 36), this comes to a total of 39,730.47 kWh (the product life can even be said to be more than 30 years).

When we now subtract the plug-in-field charges of 120 kWh per kW peak installed, which the county shall deservedly receive for the service, we are then at 39,730.47 kWh - (30*120 kWh) = 36,130.47 kWh.

Now we take the wholesale price of thin-film modules on the basis of amorphous silicon from December 2012 of 430 euros/kW peak plus 19% German VAT, adding up to 511.70 euros/kW peak. Finally we divide this amount through the 36,130.47 kWh.

As the result we get <u>1.42 cent per kWh.</u>

This is a simple calculation of course, but it shows in a comprehensible way what we can achieve with photovoltaics, even in countries that are not blessed with a lot of sunshine.

Logically, this 1.42 cents per kWh are just the amount for electricity during the day time. For our energy supply at night we ideally buy/finance our own batteries as well. For night-time electricity the amount consequently rises by the price of the battery. Both prices together, the day-time price and the night-time price, then form the final average price of our own cheap electricity. We will see a short calculatory example in the next sub-chapter.

By the way: The electricity production costs, at an even lower level of solar irradiance of 1,100 kWh per square meter and year (like in northern Germany), are also at an economic 1.77 cents/kWh. Let us apply the same calculation again: 900 kWh annual electricity production per kW peak (this is the output from 1,100 kWh solar radiation per square meter and year [Kost et al. 2012, p. 11]), plus 30% due to solar tracking technology, equals 1,170 kWh. For the period of 30 years and taking into account the degradation, this amounts to as many as 32,506.75 kWh. Minus the fee for the plug-in fields of 120 kWh/year times 30 years, we get 28,906.75 of remaining kWh. The purchase price of 511.70 euros divided by this number of kWh finally equals 1.77 cents/kWh.

Impressively favorable numbers!

Excursus: "Applicability of the Remaining Renewable Energy Forms in Comparison with Photovoltaics"

How suitable would the other renewable energy forms be? And why do we not prefer wind energy as the main source of energy instead of photovoltaics for our concept? The answer is: In comparison to wind power, which is quiet efficient as well, the electricity production costs using photovoltaics as principle technology are already lower. With wind power the kWh-production costs are currently a few cents higher, i.e. at 8.1 cents/kWh (Küchler/Meyer/Blanck 2012, p. 3), and when stored for hours with little wind they go up again by the costs of the interim storage. Generally however, wind power is the ideal complement to our photovoltaic modules as, in hours of little sunshine, usually more wind is available and the other way round.

According to the Fraunhofer Institute, wind power on the one hand offers only little additional cost reduction potential for the future, but on the other hand, the general availability of wind in most countries, just like with photovoltaics, is the key success factor here. The institute calculates that production costs for onshore wind power could go down to approximately 6-6.8 cents/kWh (Kost et al. 2012, p. 20).

Regarding photovoltaics, there is still potential for further price reductions, both for the thin-film or the crystalline technology. Thin-film photovoltaic modules sank in wholesale price during the year 2011 by 40.3% and during the year 2012 by 32.8% (pvXchange 2013) and they continue to sink. The main reasons for the decrease in price are the technological improvements and the achieved economies of scale due to more mass production. If the mass production were to be further expanded, as would happen through the elevated demand triggered by our concept, the current production cost of own electricity could – only by means of that – be further reduced. In a Fraunhofer study on the subject it is estimated: "The trend points to an approximate 20% price reduction when the cumulative installed capacity is doubled." (Wirth 2013, p. 8).

By the way, thin-film modules can work very well with diffuse light and are therefore the most price-efficient choice for solar parks in climate regions which often have rather cloudy weather conditions. Residual light can still be used by them to an extent of 96.5% if the temperatures do not exceed 25 °C (Thin-film module "TS Full SJ TS410" by the company T-Solar: *"At an irradiance of 200 W/m² and at a cell temperature of 25°C, the panel efficiency is 3.5% lower than the efficiency at Standard Test Conditions"* [Grupo T-Solar Global 2013, p. 2]).

And why are we not making use of hydropower, biogas and geothermal energy as the main energy source or at least as the complementary technology in our concept? These mentioned renewable energy sources are all sources of power, which nature is offering to us, so it makes absolute sense to use them where they are available, or to develop them where they can easily be made available. Unfortunately, the exploitation on a large scale is in my opinion not possible in such a way as it is with sun and wind at the present time and in the near future. A few explanations:

Concerning hydropower: Often hydropower is already harnessed to a high extend in most countries. The geography in the average country with only a few mountainous areas mostly impede a further and significant exploitation of it. So hydropower would not be a big solution. Also because the power production costs are at 7.6 cents/kWh (Küchler/Meyer/Blanck 2012, p. 3) without tendency for much further reduction.

If we look again at photovoltaics, we have no limitations in terms of expansion. In our market model even every tenant, who does not have a roof top or garden at his disposal, can buy and profit from photovoltaic modules. So, theoretically and practically 100% of the people in most countries can energize themselves completely with PV panels on the (still to be established) plug-in fields.

And what about biogas? In 2010 the average selling price of biogas was 8.1 cents/kWh. The pure production cost could even be said to be at only 6.2 cents/kWh. However, we would still need to add further labor costs here because of the constant need to supply biomass to the facility, which would in consequence account for higher running costs.

And geothermal energy? The electricity production costs here are very hard to number because they depend on the individual depth and complexity of the drilling on the specific site. Geothermal energy in general is an enormously good resource for electricity generation because it delivers heat to operate steam turbines day and night! So if your country disposes of geothermal heat close to the surface (for example if situated alongside the borders of tectonic plates) you should absolutely harness it. If the heat is not as easy accessible and the fracking method would be needed, it should not be harnessed in my opinion, at least not by chemical fracking (Fracking: Chemicals, which might endanger the groundwater are pumped into the deep rock in order to break it up and later extract hot water from the artificial underground cavities.). To repeat myself, geothermal energy is a very promising field, in which definitely more research should be done. Unfortunately for our concept it is not suitable as it cannot be applied in many countries due to the limitations of the current technology, whereas photovoltaics and wind power can.

To sum it up, hydro power, biogas and geothermal energy have more current limitations than photovoltaics and wind energy, plus, they remain more expensive as well in the near future. And, if we look at the flexibility concerning expansion, PV modules are impeccable. Whenever we want to increase our energy production we just buy new panels and have them plugged-in on the plug-in field. As many as we want and when we want. It is as simple as that.

The adding of new panels becomes interesting, e.g. when we want to switch our heating system from oil or gas to an electric infrared heating and, in a second step, when we want to buy an electric car and have it filled-up at our own "sunlight-gas station".

Why Batteries?

Batteries are the ideal storage addition to the photovoltaic modules for our private consumption because, as a storage possibility for the night and the twilight hours, they enable the use of the self-produced power even when the sun is not shining. Currently the prices are starting to fall because of the growing interest for solar batteries and the therefore increased mass production.

For our concept the batteries can be purchased as well at the low manufacturers' price at the service center of the county, i.e. just the bare batteries are purchased again without the need to buy any further equipment, such as control elements, an inverter and an enclosure. The batteries are then, just like the PV modules, plugged into the existing infrastructure in the plug-in solar park, inside a battery storage hall.

Let us now have a look at the battery costs because we want to know, of course, whether it is really worth using batteries as our preferred means of electricity storage for the night. To determine the price performance ratio of a battery, i.e. the amount of "cents per kWh for the electricity stored and delivered again, throughout the battery life", one must compare various characteristics of the batteries available.

You will not need to do this calculation yourself, because you will receive this information on-site at your county service center, as soon as they are established, or you will be able to find the necessary help on their website.

Below just a quick example. It is based on a genuine product offered by a German manufacturer

(Currently we see a competition mainly between less efficient but cheap lead acid-/lead gel batteries and more powerful, longer lasting but more expensive lithium-ion batteries.).

Determination of storage costs in "cents / kWh" of a lead-acid battery like it would currently be available for purchase from the manufacturer:

Assumption regarding our needs:
"In my household in the winter, from sunset until sunrise, I usually need 6.1 kWh energy."

Information provided by the manufacturer concerning the example battery:

Manufacturer price of the lead acid battery per unit including German VAT (19%)	1,309 Euro
Storage capacity	6 kWh
Efficiency (ratio of stored and re-supplied electricity)	80%
Discharge (discharge possible without damaging the battery in the long run)	50%

Intermediate Result 1:
From the above data we can deduce an actual re-delivery capacity of 6 kWh * 80% * 50% = 2.4 kWh

Intermediate Result 2:
We realize that we need three of these batteries to cover our after-sunset electricity needs of 6.1 kWh.

Let us continue the search for the storage cost "cents / kWh" (it is not important here if we consider three batteries with a correspondingly higher, summed-up kWh number or only one battery).

An additional indication, which we still need from the manufacturer is the number of possible full cycles of charges and discharges. In our case these are 3,200 (even after that the battery still has a remaining capacity of 80%, according to the manufacturer).

We therefore compute: 2.4 kWh * 3,200 = 7,680 kWh. This is hence the number of kWh we can achieve with the example battery.

Eventually, we take the price, divide it by the number of possible kWh and get our final result: 1,309 euros / 7,680 kWh = 0.1704 euros / kWh or 17.04 cents / kWh.

We note again that even if we add the cost of 1.77 cents for producing a kWh photovoltaic electricity, we are still below the current power price of many countries. So even during the light-free hours, on evenings and at night, this concept has proven to be more economic than the current system. Are you surprised now?

To obtain our average electricity price during a 24 hour period, we bring together the current cost of the battery power at night and the cost of the photovoltaic power during the day. Let us say that if we typically consume 60% of the power during the daytime and 40% in the evening / at night, then we get the following result of our average kWh price (calculated with the higher output price of northern German sunlight conditions of 1.77 cents / kWh):

(1.77 cents / kWh * 60%) + [(17.04 cents / kWh + 1.77 cents / kWh) * 40%] = <u>8.59 cents per kWh!</u>
<u>Thus, a great reduction in comparison to the current household electricity price of a lot of countries!</u>

The avalanche of PV power is getting ready to break loose. With this concept, it is possible within the coming few years! Wouldn't you agree?

If you want to find out now how many batteries and how many PV panels you need approximately, observe how many kWh you consume during the day and how many you consume at night. Note down for a few days the kWh-figure of your electricity meter at sunrise and at sunset. In this way, you will get a good average value.

Again, you will be able to carry out this calculation together with your county service center , i.e. you will at the same time have the opportunity to do so via their website.

Why the Consumption of Your Own Electricity Is So Advantageous

In the previous sections, we have learned how much cheaper it is to favor self-produced power and self-provided storage compared to the current prices of the electric supply companies.

With our own photovoltaic modules, we not only have the control over the current electricity price but also over the future power price, because we have the price of electricity for the coming years in our hands. Most forecasts predict a long term electricity price increase due to rising prices for energy resources and ongoing inflation. By consuming our self-produced energy we are therefore much better off and are no longer at the mercy of the big energy companies with their often non-transparent price increases. For our own electricity we will only pay our small, self-determined monthly rates to our national development bank, which always stay at the same low level no matter how high the price of electricity of the utility companies rises.

However, we do not just save money and have the control, we can even earn money with our generated energy. The surplus production, which you do not use for yourself can be "shared" within the county community and sold to someone, who is undergoing a power shortage at the specific moment. For example, daytime electricity could be exchanged at 10 cents / kWh and nighttime electricity at 20 cents / kWh.

A proposal regarding the process: In order for the power exchange and the payment for it to always run smoothly, the participants would need a small and simple savings account with the county, which should always be liquid and equipped in the beginning with an amount of e.g. 100 euros by the participant. From this account, the electricity donors will receive 10 cents per kWh during the day and 20 cents at night. In this way, one would not have to go back to the public electricity network with the more expensive supplier prices of the energy companies.

If at times there is no demand for interchanging power within the county community, the surplus generated goes to the electricity stock exchange, where also the best possible price shall be obtained. Self-produced electricity therefore, is also a very good financial investment! It will then be very exciting to check online via PC, tablet or mobile phone the individual level of the self-produced power, consumption of that power and the sales profits.

With your own power production and consumption, you not only save a lot in comparison to the prices of the energy companies, the cost of electricity can even be free for you if you produce surpluses and sell them. Yes, you can even turn your electricity into a net source of income. Instead of electricity bills you will receive money transfers. Sounds beneficial, does it not?

Our economic model is similarly beneficial to the state, among many things, here because it does not need to be subsidized by a feed-in tariff. This model finances itself and stands on its own two feet (if your country already has granted feed-in tariffs, you will read later on how our concept can bring down the future costs of these guaranteed long term payments).

However, it is not only the purely economic arguments that make us want to go from the centralized, far away organized electricity production, towards power produced in our own region, generated by us and our neighbors all around (this model remains centralized only in the sense that the county administration manages the infrastructure and offers support). Further, it just feels good to be a high-tech ecologist – if you will. I would say, this is what can be called more than being up-to-date.

Consumption of self-produced electricity adds, so to speak, not only to our budget but also to our pride. After all, we automatically get to be a positive example to our neighbors here and in the whole world.

With the described county-plug-in model, we do not only achieve a 100% secure energy supply perspective for all time, we also begin to "cure" our environment from what our current energy supply has caused in terms of "injuries". Self-interest and usefulness to the public is thus connected in this concept in the best of ways.

Why Is the New Idea of the Plug-in Fields So Beneficial?

The first benefit is that when we read the name "plug-in field", we immediately have a good picture in our mind of how the system works. It is a simple principle, which is easy to understand.

The most important advantage however, is that the infrastructure of the plug-in fields provided by the county reduces our photovoltaic power price by about half compared to an installation on the roof! For our application we only need the bare photovoltaic modules and therefore we save ourselves the costs of the mounting frame, of the control devices, of the solar installer plus the costs of maintenance and care (*"The price of the PV modules is responsible for a bit over half of the capital cost of a PV power plant"* [Wirth 2013, p. 7], or even for less, depending on whether the specific installer adds a profit margin to the modules). With the plug-in fields provided by the county, an important prerequisite for our low electricity price is therefore established.

Very beneficial in our concept of the plug-in fields is also the fact that one does not need to be a homeowner with a private, optimally aligned roof in order to benefit from the power of photovoltaics. All tenants, apartment owners and homeowners – and completely irrelevant to whether they live in the city or in the countryside, in a basement apartment or in a skyscraper – can produce their own electricity for the household, heating and an electric car! The electricity comes from their modules and batteries to their apartment- or house electricity meter, so to speak. (If we assume that electric cars will have an individual electricity meter as well, they could also obtain their own plug-in power, no matter where in the county they will be parking.)

As another positive aspect, the plug-in concept allows an overall very simple participation process, which we can imagine like this: First, the participating citizens select the best photovoltaic panels and batteries in the service center according to the best price-performance ratio. Then the panels and the batteries are mounted by the district's electricians on the plug-in field. After that, they are registered with a few clicks on the net. Done! From that moment on, the participants are able to conveniently graphically follow, on their tablet PC or on their smartphone, how their photovoltaic panels (and wind turbine shares) successfully produce power.

Let's come back again to the phrase "virtually free of charge" from the bold text at the very beginning of the first chapter. Reading that you might have already deduced correctly that the county should be rewarded in some way for its service. Now, these fees are not payable in money but, more practicable, in another currency – in kWh of course. And here as well we come across a positive surprise: The fees of the 10 kWh per kW peak of installed modules are being more than recouped for the participants through the technique of the automatic sunlight tracking, which the plug-in fields will be equipped with. The PV modules will be able to carry out a two-axis movement and in this way follow the path of the sun in an optimal manner. This technique, which would be very expensive on a roof, increases the electricity yield, as already mentioned, by up to 30%. The fees of the 10 kWh / kW peak in contrast account for less than 10% of the yield. Thanks to the county and thanks to the solar tracking we thus generate about 20% more net yield. On top of that, during winter, county employees will free the modules daily from layers of snow (as necessary) with a snow blower, so that the electricity can always flow at maximum capacity. A nice luxury that the roof-system operators unfortunately do not usually have.

One advantage of plug-in fields in terms of visibility in the landscape is that bushes can be planted around them and in this way, a large photovoltaic park in a plain area is not necessarily noticeable, i.e. the natural appearance of the landscape remains intact. Or, as another option, one leaves the view of the photovoltaic modules, similar to fields of greenhouses, and even enjoys this view of environmentally friendly energy production. So, it can be done just as desired by the respective county community.

Why Is the County Predestined to Play a Central Role?

The counties, as organizers of the plug-in fields for their citizens, make a lot of sense. They are both, the central administration point, i.e. the long arm of the state, and the community network of the people at the local and regional level, with which the residents identify and feel connected.

Our concept most probably creates an even stronger identification when the district proclaims: *"Your county now offers the possibility to supply yourself with your own, cheaper electricity."*

The counties with their county seats are also large enough to offer the necessary infrastructure and to administer it, and they are not too big and not too far away, but in the true sense of the word close to the citizen. They can concentrate competence and act in a more economically viable way than the individual communities could. And, they already have experience with regional and local energy supply through their municipal utilities.

The self-owned photovoltaic system supervised by the county, is thus a mix in the wider sense of "central" and "decentralized" energy supply, and therefore appeals to the sense of security of both mentalities of the citizens: the one mentality-type citizens that can not yet imagine quite how they could supply themselves with electricity using their own system, just out of the habit of centralized energy supply over the recent decades, and the other mentality-type citizens, who feel safe and comfortable with their own photovoltaic modules because they can finally produce cheap, clean and safe power for their own consumption. The county will take care of everything technical, it will monitor the systems and will fix all possible faults directly. The citizen can thus always have a well-attended and safe feeling.

As the people in most countries live at the same time in a county (if the country is organized in this way), they will also quickly feel involved at a high level when their districts offer to implement this concept for their residents, i.e. for us. When the people know that the PV fields are professionally managed with the expertise of the counties, the motivation for active participation of the citizens will be high, in my opinion. Through the confidence that the county government enjoys, the people will invest even more in clean solar energy, as they have been already doing in roof-top systems and other solar plants.

More advantages of the county as the organizer of the photovoltaic fields:

All districts of the country together can and should act as a single customer towards the manufacturers of PV modules, batteries, control devices and mounting systems. In this way, an extremely high level of buying power can be achieved. The purchase discount will therefore be as large as possible, which will thus account for an enormous cost advantage to be passed on fully to the citizens.

Obviously each county can also individually buy, mounting frames e.g., from the industry in its own region. In this way, the local economy can be supported. The cost-benefit decision can remain, in this case, with the county as long as the same nationwide quality standards are maintained by the manufacturers. With this, the strengthening of the local economy and more motivation for innovation can be achieved. Depending on the quality of the produced items and the demand for them, a new national or even international industry branch could be created in specific counties that way.

Another reason for each county as the best place for the plug-in fields and the wind turbines is that the energy production throughout the country will then be geographically so equally distributed that long-distance power lines will not be needed anymore to the same extent as before. Billions of euros could be saved for planned long-distance lines which would consequently become unnecessary. Billions of euros, which can then remain in the pockets of the citizens.

One challenge is of course the demand for land. In case a county seat does not have enough space for one or more plug-in fields, the surrounding communities will surely help them out. The reason being: Their citizens will benefit as well from cleaner air and from the lower electricity prices in the county.

Although the counties will be responsible as the local organizers, they will of course be supported by unified national administrative standards and, e.g., by the necessary software tools. In this way, the power exchange mechanism between the participants will hence be automatically controlled by special software. The transmission of the surplus power to the supra-regional cooperatives, for the purpose of selling it at the electricity stock exchange, will also happen automatically. This means that the counties do not need extra personnel for this and the workload can be reduced.

The administrative fee of, e.g., 10 kWh/kW peak per month and per participant, which the county may charge, can be used for the energy supply of the county itself or it can be sold to the other county communities. In case there is more energy available than needed at a specific time, the county can interchange/sell it among the participants of the plug-in fields as well, or, of course, sell it at the electricity stock exchange. Regarding the interchanging of power, the county shall be given priority over the private households in delivering the power, to ensure that the county is always able to cover the costs incurred. Thus, the county supports its plug-in-field participants and the participants support the county. This can be called a true symbiosis.

Why Finance?

If you have your own capital at hand, you can achieve the outstanding "profit by savings", calculated above, when investing it in your own photovoltaic panels and at the same time supply yourself for 30 years and longer with your own power. If you do not have the capital for all PV modules and batteries at hand, financing is an equally good choice. After all, you only pay a 0.5-1% interest rate to your lending bank, i.e. to your national development bank.

We learned that PV modules are much cheaper for private electricity consumption than the current price of the electricity suppliers. How little your monthly payments will be exactly depends on the height of the rates that you yourself determine. The question will be: "Will I save half or even two-thirds of my previous monthly electricity price?" Since photovoltaic modules have a lifetime of over 30 years, you can freely distribute the payments over this period, or pay the full amount at any time (without any payback-fees of course). So, with this concept you can completely determine your monthly electricity price yourself!

Why Pan-regional Cooperatives for Sale and for Temporary Storage of the Plug-in-field Power?

In order for us private producers to achieve a good price on the electricity stock exchange, it is important that the interests of many plug-in-field participants are bundled. A supra-regional association or cooperative can accomplish that because of its size and the according weight it would have on the electricity market. Such a cooperative shall be legally obliged to work absolutely transparently and to always publish business processes and numbers on the Internet. Similarly, there shall be a zero-tolerance rule concerning lobbying influences, which would need regular verification by auditors.

Due to its size, such an interest group is also predestined to manage the all-important high-volume intermediate storage of the electricity produced during the day for sale at night. This intermediate storage, in national or international pump storage plants for the evening and night hours, is elementary to stabilize the power supply of the plug-in fields and thus making the photovoltaic electricity a base load power. The cost of storage for the night should be integrated into the price for the day-time power so that the electricity at night does not need to be expensive (i.e. does not exceed the average market price of electricity) and so that at the same time, the daytime electricity will not become too cheap (does not go to zero in the extreme case).

The long-term storage for the winter, which is also managed by the cooperatives, should ideally be done via Power-to-Gas power plants (in which the citizens should be able to participate financially) and the gas should be stored in the national gas grid.

Excursus: „Electricity Storage"

As for the large-volume energy storage, currently two types, or technologies are practical. First, the pumped storage hydro power stations and second the emerging Power-to-Gas technology.

Depending on the geography and availability of water in your country, you might have more or less pump storage capacity. If your country does not have enough capacity to cover your needs, maybe your surrounding neighbors do. The pump storage technology has a degree of energy efficiency of about 70% (Wirth 2013, p. 56) and it is well-proven. If you know the pump storage capacity of your country in GWh and if you assume that a three person household has a demand of 3.5 kWh per night, you can calculate how many of these households can be supplied with your country's capacity.

An interesting concept for the development of new pump storage capacities in Germany is to store water within the towers of wind turbines. A pilot project with four towers and a pump capacity of 70 MWh (12 MW rated power) is currently being planned near the city of Gaildorf (Municipality of Gaildorf 2013).

Power-to-Gas Capacity: The Power-to-Gas technology, as its name implies, allows the conversion of electric power into gas (hydrogen or methane). This is accomplished by electrolysis. In many countries a huge gas pipeline grid already exists, which offers an enormous storage capacity. Such a network of pipelines makes the Power-to-Gas technology a practicable and attractive option for large scale energy storage. The gas can be transported via the gas grid or, as just mentioned, stored and at any time to be converted back into electricity. Depending on the size of the grid in your country, up to several months of the electricity demand can be stored in it. In Germany for example, the storage reservoir is more than 200 terawatt hours (Fraunhofer-Gesellschaft 2010).

The efficiencies in the conversion of electricity-to-gas are currently at a maximum of 77%, regarding the conversion electricity-to-gas-to-electricity, they are currently at a maximum of 44% (Sterner et al. 2012, p. 16).

With the two described storage technologies and with our photovoltaic power, which we complement with wind power, we can fully supply ourselves with all the energy we need. However, the old, conventional power plant landscape does not necessarily need to remain completely unused. By conversion of the existing power plants to Power-to-Gas plants, a good portion of the useful infrastructure can remain.

Strategy for the Winter

In case you have a big weather difference between summer and winter in your country, the following strategy will bring your country through the dark season. Let us take Germany again as a concrete example. While the solar radiation in Germany from the spring until the beginning of autumn is encouragingly high, both in the South and in the North (which means that the people therefore can supply their entire household including electric car with comparably few photovoltaic modules), the solar radiation quantity in winter, with its short and frequently cloudy days, is unfortunately relatively small.

To illustrate this, let's have a look at three decades of averaged values of solar radiation in Germany.

Nationwide averages of monthly totals (1981 - 2010) in kWh / m²:

	Maximum (Rather South and North East Germany)	Medium	Minimal (Rather Low Mountain Range and North West Germany)
January	50	23	15
February	77	40	30
March	115	75	66
April	127	117	107
May	168	153	139
June	175	159	143
July	180	161	145
August	153	137	124
September	107	91	80
October	76	56	46
November	45	25	19
December	38	17	11

(Data: National Meteorological Service of Germany 2012)

You will find that in the sun-rich regions in Germany (column "Maximum"), the highest value in July with 180 kWh / m² is 4.7 times the value of December with 38 kWh / m². In the less sunny regions (column "Minimum"), the ratio is even greater.

To illustrate, here are the above figures shown as a graph:

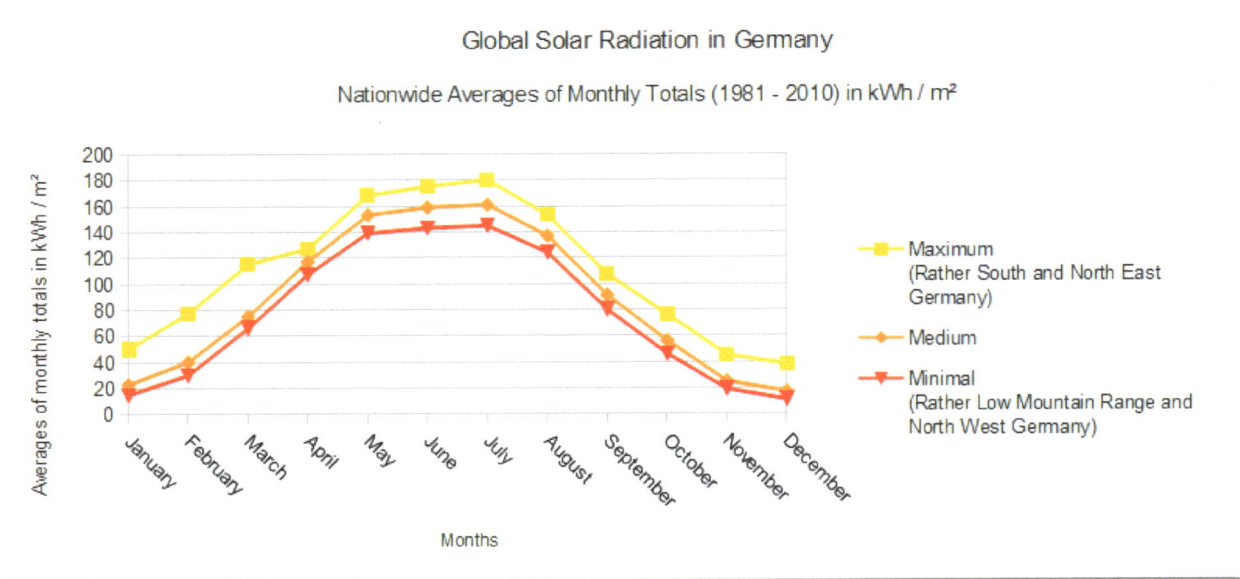

Global Solar Radiation in Germany

Nationwide Averages of Monthly Totals (1981 - 2010) in kWh / m²

(Illustration: own diagram; data: National Meteorological Service of Germany 2012)

To have a practical application, let us take the example of a three-person family, which we will come across more often in this book. Let us assume that this family covers everything including heating and electric car with their self-produced electricity and let us estimate the following, as we shall see later, realistic values:

Electricity for lighting, appliances, etc.	10 kWh per day
Space heating (5 months: November to March)	33 kWh for a winter day
Hot water production	5 kWh per day
E-Car	8 kWh per day

The total power requirement for a day in December (including space heating) in the example is 56 kWh. If we wanted to produce this amount of electricity using purely photovoltaic on the same day in December, we would need the huge amount of photovoltaic modules of 99.5 kW peak (This calculation is based on the average solar radiation conditions specified in the "Medium" column in the table above).

On a day in July, the need for electricity is at 23 kWh, i.e. 33 kWh less than in winter because no space heating is used. For this 23 kWh, a module quantity of only 4.16 kW peak would be needed, of course due to the better radiation conditions during the summer as well. This is a pleasantly small number. Just for comparison: At the same high consumption level as on a winter day, that is at 56 kWh, the family, in July, would also only need a set of modules of few 10.14 kW peak.

The difference from the extreme value of 99.5 kW peak of PV modules is enormous and it becomes clear that it makes more sense to equip ourselves, say, only for the months of March to September with enough photovoltaic panels and for the remaining months to benefit from an additional energy source. But we will get to that later.

First, the calculation of the necessary module capacity for our example family if they already want to fully supply themselves daily in March with solar power at the solar radiation of 75 kWh / m² / month (i.e. at 2.42 kWh / m² / day) (calculated for the higher demand of 56 kWh):

56 kWh / [2.42 kWh / m² * (900 kWh / 1,100 kWh / m² "photovoltaic power efficiency ratio of Northern Germany") + 30% more yield through solar tracking] = a still reasonable amount of PV modules, namely 21.76 kW peak.

With this amount of modules, the family, of course, produces a lot more electricity over the summer than it needs. Instead of selling the power directly, and now comes a keystone of the strategy for the winter, the plug-in-field participant shall be able to save up the energy for the winter months – with the Power-to-Gas method.

Since the efficiency of the conversion of electricity-to-gas-to-electricity is currently 44%, a little more than twice the winter shortfall of kilowatt hours would have to be saved up in the summer. With the mentioned PV module quantity this will not be a problem for our example-family. It will produce more than it needs to save up for the winter and will be able to sell the rest (after payment of the fees of 120 kWh / kW peak and year) at a profit.

The following table shows the amounts of generated kilowatt hours. You will notice how much more energy than needed is produced on a summer day, respectively in a summer month:

	Solar Radiation ("Medium" Column) in kWh / m² Per Month	Electricity Production Per Day with a Module Quantity of 21.76 kW Peak (in kWh)	Consumption Per Day (kWh)	Surplus of Electricity, Respectively Lack of Electricity (When "-") Per Day (kWh)	Surplus of Electricity, Respectively Lack of Electricity Per Month (kWh)
January	23	17.17	56	-38.83	-1203.63
February	40	31.93	56	-24.07	-698.13
March	75	56	56	0	0
April	117	90.27	23	67.27	2018.16
May	153	114.24	23	91.24	2828.44
June	159	122.68	23	99.68	2990.32
July	161	120.21	23	97.21	3013.61
August	137	102.29	23	79.29	2458.09
September	91	70.21	23	47.21	1416.35
October	56	41.81	23	18.81	583.21
November	25	19.29	56	-36.71	-1101.33
December	17	12.69	56	-43.31	-1342.51

(Data: own calculation, National Meteorological Service of Germany 2012)

Logically, if the household for example uses a gas heater or a gas stove, the surplus of kilowatt hours can be accumulated in gas and without the conversion back to electricity.

The save-up for the winter could be organized like this: The participants have a "kilowatt hours or gas storage account", similar to a bank savings account, with their supra-regional cooperative and for a small fee (again in kilowatt hours) the cooperative administers this savings accounts.

The saving of course works only when sufficient Power-to-Gas power plants have been built. Until then, <u>in order to bridge the time gap during the construction period of new power plants and to relieve the gas network and our night storage batteries, we need, as mentioned above, a complementary energy source to photovoltaics.</u> A source that is available and applicable everywhere, and above all, delivers energy in an intensified manner preferably at the times when the sun is shining less. From experience we know about the frequent correlation between "little sunshine is equal to more wind". With wind power as the complementary form of energy in our approach, we can make up for the disadvantage of the "bad" autumn and winter weather, and even use it to our advantage.

In the following illustration, regarding the monthly wind power production in 2012 in Germany, we can see when wind power is especially available.

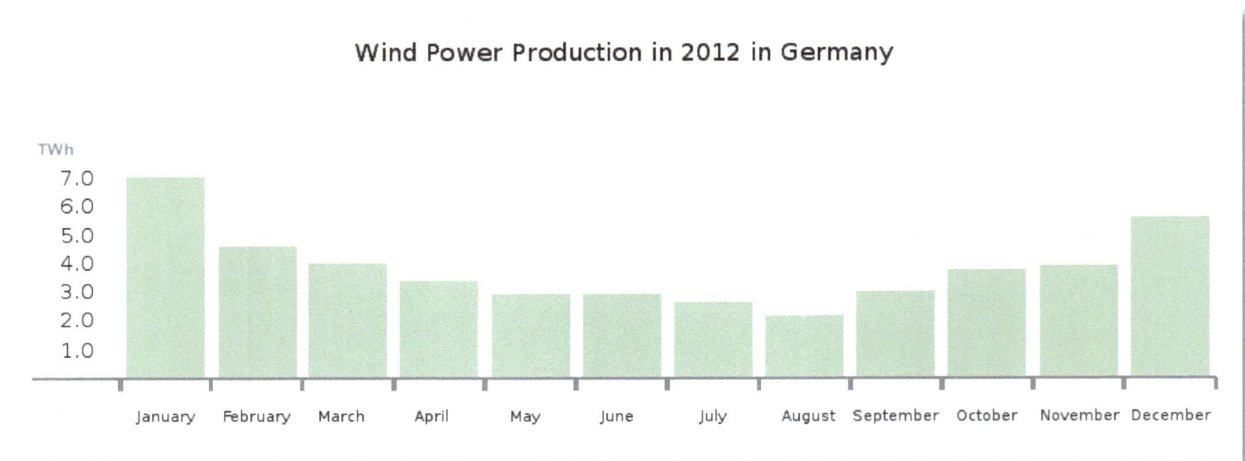

Wind Power Production in 2012 in Germany

TWh
7.0
6.0
5.0
4.0
3.0
2.0
1.0

January February March April May June July August September October November December

(Illustration: Burger 2013, p. 14)

We notice that wind power in winter – but at night as well – can also provide us with cheap electricity. Not at 1.77 cents / kWh, but therefore, as already mentioned, at 8.1 cents / kWh. When the wind blows at night, these 8.1 cents are of course cheaper than our electricity stored in batteries, which we would not use up then.

As for the procedure, it could take place in the following way: The county has a wind turbine constructed once it has a critical mass of interested participants, e.g. for 30% of the shares of the wind turbine. The interested citizens purchase the shares of the wind turbine from the county and then, depending on the amount of the shares, the owners receive the appropriate percentage of the energy yield. The remaining shares are then gradually sold by the county and in the interim the county uses the electricity produced for itself. When all the shares of a wind turbine are sold, another wind turbine can be built as soon as the minimum number of interested participants is reached again.

The turbines could then function simultaneously as pump storage power plants, which would make such a wind tower promptly even more profitable, especially if you look at the power exchange rate at night of 20 cents / kWh.

Concerning the future distribution of wind power plants in the world from the point of view of our concept, it is likely that they are in greater demand in the further northern countries in order to compensate the lower solar radiation there as necessary.

As for the popularity, with respect to aversion regarding the wind turbines, it can be assumed that when the citizens are at the same time the owners of the windmills, they are more likely to accept that the turbines are visible elements in the landscape. Perhaps, they will even actually be very proud when pointing at them.

The use of wind power would, advantageously, then also reduce the need for land for the photovoltaic plug-in fields.

This Is What We Can Leave Behind – Almost

On the way to a "better" energy we can simultaneously leave all dirty energy sources behind us – almost – because we still need to deal with the unhealthy legacy of some forms of energy production from the recent decades far into the future.

On the one hand we have CO_2 causing fuels such as heating oil, gasoline, stone coal and brown coal, which were and are still simply burned with a very low efficiency. At least regarding these energy sources we still have the chance to undo their harmful inheritance in a manageable number of future generations. However, currently they still pollute the air we breathe and continue to heat up the climate without the initiation of any real turnaround.

Then there is nuclear power. The thoughts of nuclear incidents caused by humans or by technical failures or earthquakes and the consequential following radioactive pollution and contamination, can often be repressed, respectively one is constantly hoping that nothing will happen. Looking at the catastrophe in Fukushima in Japan, we now are however very conscious that even industrialized high-tech nations can be hit. Yet, what is happening continuously and everywhere in terms of radioactive contamination, is the "controlled accident" – the nuclear waste produced! As we all know, this waste will radiate several hundreds of thousands of years, if highly radioactive.

Besides the ever-multiplying health risks of conventional energy, one does not even like to talk about financial costs but of course, they are also there. Here we talk about costs that are not included in the electricity price, but must nevertheless be borne by the society, the so-called "society costs" or "eternity costs". Don't you agree that if the nuclear companies would have been obliged to bear these costs, they would have abandoned the nuclear energy market many years ago on their own?

The following are the additional costs on top of the household energy prices, borne by the tax paying society as a whole. The numbers are taken from a German study commissioned by the Greenpeace Energy eG and the Federal Wind Energy Association (Bundesverband WindEnergie e.V. - BWE). However the situation can be assumed to be similar in all countries using conventional energy sources: For each kWh of nuclear power the society pays, according to the study, again on top of the average household electricity price of 27.3 cents / kWh, between 36.8 cents / kWh and 11 cents / kWh in the form of non-earmarked taxes or insurance premiums – depending on whether a major incident such as in Japan occurs or not. In the case of power from stone coal it is 9.4 cents / kWh in addition, for brown coal it is 10.2 cents / kWh and for natural gas it is 3.6 cents / kWh extra on top of the price of electricity (Küchler/Meyer/Blanck 2012, p. 11). So if energy forms deserve the criticism of "not being affordable" they are the conventional energy forms.

The "wagon train" of the dirty forms of energy, which is still going at full speed, must be stopped quickly. Every year in Germany there are about 450 tons of radioactive nuclear waste produced. Worldwide, there are about 12,000 tons per year (World Nuclear Association 2012). For CO_2, the emissions in Germany per year are at a difficult to imagine 803 million tons, the total of the world is at 34 billion tons in the same period (Federal Ministry of Economics and Technology 2013, p. 12). Each year earlier that we can switch to a better form of energy production, we save these quantities of contamination and pollution and thereby prevent a permanent increase of more and more unhealthy burdens. So let us do that for us, for our health and for the health of our children!

Let us protect our drinking water and our air! And let us observe if we succeed again in reducing the occurred respiratory diseases and the cancer rate by doing so! With photovoltaic plug-in fields all over the world, in each regional district, we could succeed.

Interim Conclusion

With the photovoltaic plug-in fields, power will become cheaper again and we can protect our health with it in the long run! The technology is there and it is economical enough. The will of the people for clean and cost-saving energy is also there. The county seats with their competence and their land are available and the development banks of your country have probably already begun promoting renewable energy projects.

We now also know how many photovoltaic modules each person needs for his electricity consumption, and even in the case that someone has his entire home fully heated with his own green power, and on top of that is driving his own electric car. How much exactly the electric heating saves, plus how economical it is precisely to drive 100km with one's electric car, we will learn later on in the book.

So now we have gotten to know the supply concept with which we can achieve so much and we realize that the only thing that has been missing was how to bring together the individual technology components and the different interest groups. To satisfy the demand for clean and cheaper electricity (possibly) from now on, thus, in the end, was only a question of how to shape the service of making the existing technology available.

In the following chapter you will learn about more advantages of this new energy plan and towards the end of the book you will read, in a suggestion guide, how it can be implemented exactly. If you then think that the guide is not complete, write down your ideas quickly and discuss them on Facebook in the international Facebook community group "Photovoltaic_Plug-in_Fields_international" under the link address:
https://www.facebook.com/groups/photovoltaicpluginfields.international/

In order to coordinate the activities around the photovoltaic plug-in fields specifically for your country, feel free to initiate a Facebook community group especially for your nation (in case it doesn't already exist). For reasons of uniformity and in order to be better findable by all interested parties, you might want to name it according to the following proposal and of course in your specific language: Photovoltaic_Plug-in_Fields_"Your_Country".

Chapter 2

Exactly what Advantages Do the Photovoltaic Plug-in Fields Offer You?

We have already mentioned many advantages. On the following pages there are even more to come and specifically for you! These advantages are subdivided depending on whether you are a private decision maker and user, a local government policy maker or a politician on the national basis.

Benefits For You as the User of the Photovoltaic Plug-in Fields!

If you liked the unexpected moments from the previous chapter, then I am happy to promise you even more in this one. You can always apply the following observations and calculations to your household, even if you live in a home that consists of less or more people than in our model example.

In the "Renewable Energies Agency" chart below, you can see the <u>development of the monthly energy costs (electricity, heating and gasoline costs) for a three-person household from the year 2000 until the year 2012.</u> The largest cost increases since 2000 that the household has had to accept are for heating oil. The household uses fossil- and nuclear energy forms as currently most private households still do. Again, the energy prices that are used in this chart are the prices in Germany. Please do some research on the Internet to find the energy price development for your country over the last few years. Most likely, the prices you have had to pay developed in the same way as those in the chart below – going up by a high percentage!

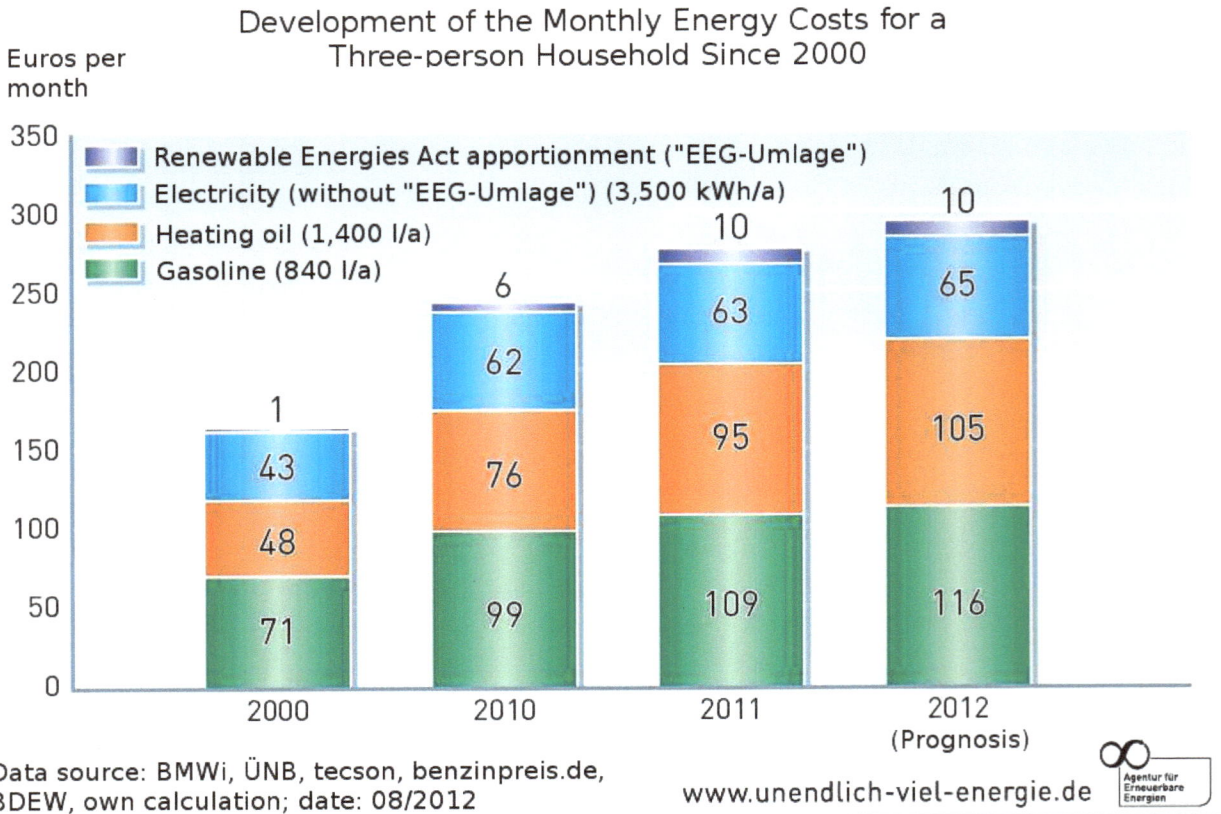

Development of the Monthly Energy Costs for a Three-person Household Since 2000

Euros per month

Legend:
- Renewable Energies Act apportionment ("EEG-Umlage")
- Electricity (without "EEG-Umlage") (3,500 kWh/a)
- Heating oil (1,400 l/a)
- Gasoline (840 l/a)

2000: Gasoline 71, Heating oil 48, Electricity 43, EEG 1
2010: Gasoline 99, Heating oil 76, Electricity 62, EEG 6
2011: Gasoline 109, Heating oil 95, Electricity 63, EEG 10
2012 (Prognosis): Gasoline 116, Heating oil 105, Electricity 65, EEG 10

Data source: BMWi, ÜNB, tecson, benzinpreis.de, BDEW, own calculation; date: 08/2012

www.unendlich-viel-energie.de

Agentur für Erneuerbare Energien

(Illustration: Renewable Energies Agency 2012d)

Let us now calculate how much lower the costs for this private household would be if the photovoltaic plug-in fields already existed today. First let us see how much would be saved regarding electricity expenses and then for the case that the household replaced its oil heating by an electric infrared heater. Additionally, we answer the question: How much can be saved if the gasoline car is replaced with a new electric car?

Electricity Cost Savings

Let us calculate what the three-person example household would save in electricity costs if it already had a plug-in field provided in its district.

Reminder:

In the first chapter we showed that power from the plug-in field during the day would cost 1.77 cents / kWh and during the night 17.04 + 1.77 cents / kWh for a north German district. We then calculated, using the example household and assuming a day-night consumption ratio of 60:40, that the electricity for this household on average would cost 8.59 cents / kWh.

Now let's go back to the monthly electricity cost savings of our example household:

In 2012, this household paid 75 euros for a monthly usage of 291.67 kWh (3,500 kWh divided by

12 months), with an average electricity price of 25.71 cents / kWh.

With their own photovoltaic modules and batteries, according to our plug-in concept, they would only pay 25.05 euros per month (291.67 kWh * 8.59 cents / kWh)! Saving: 49.95 euros per month!

Depending on how much excess capacity of PV power the family has, it can exchange the surplus electricity, as mentioned, at 10 cents / kWh during the day, and respectively at 20 cents / kWh at night. Or, it can sell the surplus energy at the electricity stock exchange (European average of the electricity market price of the leading electricity stock exchanges from 01/2008-10/2012: 5.3 cents / kWh [Küchler/Litz 2013, p. 4]). Hence, it is very well possible that the family could earn an additional income in the summer by doing this.

Heating Cost Savings

What do the big expenditures for heating and hot water look like for the three-person example household?

Concerning the expenditures for heating oil, let's look at whether the example household can save money with an electric infrared heater in comparison to an oil heater and if this is possible, how much they can save.

First we separate space heating from the domestic hot water. The national average share of domestic hot water in relation to the total heat energy costs equals 13.46% (in Germany, year 2010) (Federal Ministry of Economics and Technology 2013, p. 7). Let us deal with the domestic hot water a bit later and first focus on the lion's share, the 86.54% for space heating. For this calculation, we use the monthly averages from the diagram above "Development of the monthly energy costs for a three-person household since 2000".

Expenditures Per Month for Space Heating:

105 euros * 86.54% = 90.87 euros / month are spent for space heating (equivalent to 100.96 liters of heating oil, in kilograms of fuel: 86.83 kg).

In a study of the German University of Kaiserslautern it was observed that electric infrared heaters generally use less energy than oil or gas fueled hot water radiators, which heat up the air in order to heat up a room (called convection heating).

In this study, the infrared heating was compared to a gas heating. In the conducted experiment, which had a time frame of several months, it was found that the gas heating with 187.85 kWh / m² (calculated on the basis of the condensing heating technology) used much more energy to produce the same thermal comfort feeling than the infrared heater with an energy consumption of only 71.21 kWh / m² (Kosack 2009, p. 34). Thus, the infrared heater requires only 37.9% of the energy that the gas heater requires!

Now we apply the study to our sample household: Since, according to the "Institute of Heat and Oil Technology" gas and oil heating systems can be considered equally efficient (Institute of Heat and Oil Technology 2013), we can use this 37.9%-ratio for our three-person sample household as well, despite the oil heating system that it has. So that means, instead of the energy from the following equation valid for the conventional oil heating system: 86.83 kg heating oil * 11.9 kWh (heat value) * 1.06 (to get the higher fuel value of the condensing heating technology) = 1,095.29 kWh, the example household only needs 37.9% of that, i.e. 415.12 kWh, thanks to the infrared heater. And, thanks to our own, cheap plug-in photovoltaic electricity, the example family would thus again pay less for their heating, namely <u>only 35.66 euros per month</u> (415.12 kWh * 8.59 cents / kWh)! This is a <u>saving of 55.21 euros, per month!</u>

The infrared heater, however, is not only more convenient concerning ongoing operation, but also concerning acquisition. According to the comparisons of the energy forum of the state of Hessen, "Energieforum-Hessen.de", the infrared radiator equipment is 25,000 euros lower in price than an equivalent pellet heating system (Energieforum-Hessen.de 2013) (approximate amount for a one to two family home)!

We now have seen that there are good reasons for opting for an infrared heater especially when the purchase of a new heating system is imminent in any case, but also if you just want to replace the expensive and dirty fuels of the past. At the moment there is already a subtle trend towards this new type of heating because, apart from the financial benefits, it makes a lot of sense for other reasons as well.

Further to the above, the Energieforum-Hessen.de points out additional, very positive features of the infrared radiators:

> *"Another plus is that the radiant heat from infrared heaters – like the radiant heat of the sun – is perceived by many people to be very comfortable, as there is no air and dust swirl like with the conventional convection heaters."*

> *"Another key advantage of infrared heating is the easy installation. While it is necessary to install a complete pipe circuit for the water with a conventional heating system of oil or wood, this is not necessary with infrared heating."*

> *"Infrared heating can thus be set up always and anywhere without any problems."* [Even without tradespeople. Author's note.]

> *"The heating produces no emissions, which must be cleaned and diverted from the living areas. So you do not even need a chimney anymore and also the cost for the chimney sweep and the prescribed emission inspection belong to the past."*

> *"Design solutions are ideal for infrared heating. You can use the heating elements for example as a mirror or as a picture element on the wall, or you can even attach them to the ceiling. Here, materials such as granite, glass or marble can also come into play."*

> *"Mirror heaters can be easily used in the bathroom. The big advantage is that the heater is integrated in the mirror – a space-saving solution that gives you a clear view immediately, because the mirror cannot steam up."*

And here an illustration of the functionality. The radiation of heat is comparable to the heating by the sun:

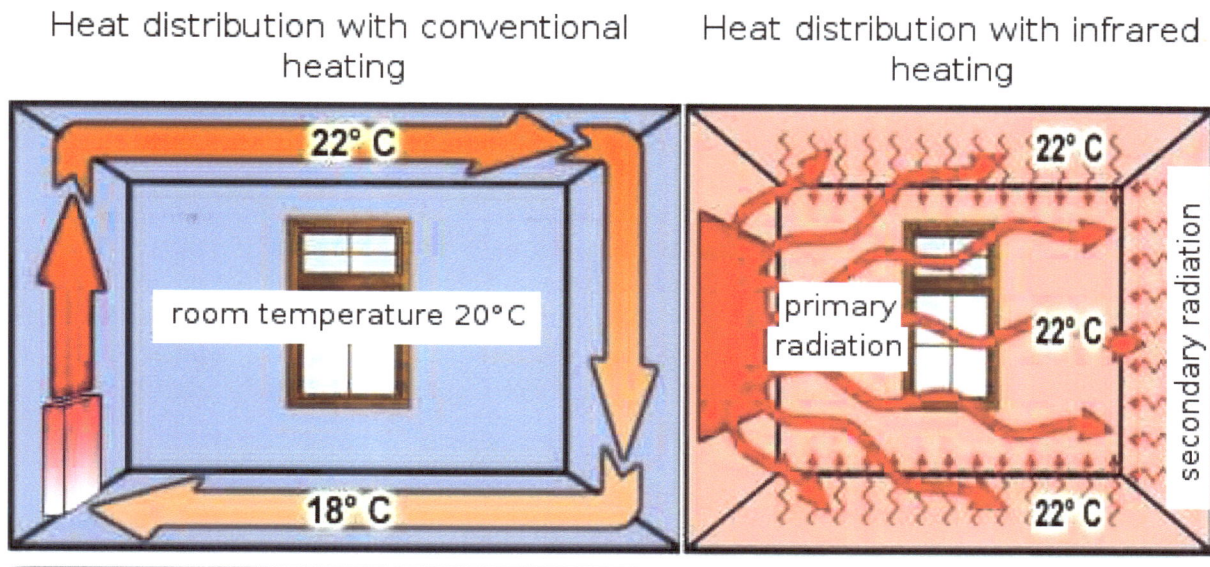

Heat distribution with conventional heating — Heat distribution with infrared heating

(Illustration: Energieforum-Hessen.de 2013)

As you can see from the illustration, with infrared heated rooms it is unlikely that one has an unnecessarily large warm air cushion at the ceiling and at the same time cold feet on the ground, as is often the case with convection heating.

Further to the positive effects already mentioned, there are still more:

- Conventional oil and gas heating systems belong to the largest power consumers in the household because of their water pump and their burner. If these heating systems are replaced, the saved power can be converted directly into infrared heat.
- Homeowners would get two rooms of their house back. The space for the oil tank and the boiler room. With the appropriate renovation, size and incidence of light, and depending on the overall quality, the "new" rooms could even be rented out.
- The infrared technology heats up a room instantly: When you come home, irrelevant of how long the apartment has not been heated, you immediately get a warm feeling – right at the push of a button. So it is no longer necessary to have the heating running when you are not at home.
- For people with asthma, the fresh, slightly cooler air with simultaneous cozy warmth and without dust swirls is also very pleasant with this type of heating.
- In addition to the previous point, the study of the University of Kaiserslautern showed that with infrared heating, the drying of damp walls is promoted and the formation of mold and all the associated health problems is counteracted (Kosack 2009, p. 39).

Due to its many advantages, the infrared heater could become a very strong trend. Together with the sustainable and cheap electricity from our own photovoltaic panels and our wind power shares, it could even become the dominant way of heating in the future.

A beneficial supplement to the infrared heating system would be if additional electric night-storage heaters would be used. Opposite to their traditional way of functioning, they would store cheap photovoltaic electricity during the day in the form of heat in order to release the heat again in the evening and at night. Since both, infrared heaters and "day-storage heating", are much cheaper than purchasing an oil, gas or pellet central heating, a combination of the two electric versions could be quite practical (the day-storage heating could also be electrical floor heating).

<u>What Can Be Achieved for the Family Concerning the Expenditures for Domestic Hot Water Production?</u>

If we assume that an oil fired boiler and an electric boiler have about the same efficiency, then a kWh of fuel oil still costs about 9 cents and our photovoltaic electricity during the day only costs 1.77 cents / kWh! Because of this difference in costs, it is of course better to use an electric boiler in our concept. The boiler is then heated during the day and through the insulation, similar to a thermos, a high proportion of heat can be stored for the evening hours and for the night.

For our three-person sample household, this means:

Previous expenses for domestic hot water production: 105 euros (total monthly expenditure on heating oil) * 13.46% (percentage of hot water production) = 14.13 euros.

To compare the costs, we now need the number of kWh of the fuel oil used. Here the lower heating value applies.

13.46% of the monthly spent fuel oil is equivalent to 15.7 liters or 13.5 kg. We calculate 13.5 kg * 11.9 kWh (to get the heating value) = 160.71 kWh.

In terms of costs for the use of the electric boiler, we therefore get: 160.71 kWh * 1.77 cents / kWh = <u>2.84 euros!</u> So, the family <u>additionally saves 11.29 euros per month!</u>

Gasoline Cost Savings

Would you have thought that the costs of an electric car like the Nissan Leaf in consumption per 100 km, even at a high household electricity price of 27.3 cents / kWh, are already at only 5.49 euros? (Consumption per 100 km: 20.1 kWh [Houben 2012, p. 2])

You surely already suspect that also here the example family is able to save a lot of money. Let us look at the current gasoline expenses in detail:

On average 70 liters of gasoline for 116 euros are needed per month (price per liter in the diagram: 1.66 euro).

How many kilometers can you drive with this amount of liters with a gas car that is comparable in size to the Nissan Leaf? A comparable car would be a frugal type of VW Golf with a fuel consumption of 5.5 liters per 100 km. The result: For 116 euros, the family could drive a total distance of 1,273 km with the VW Golf.

For this range the electric Nissan Leaf requires 255.82 kWh. These cost, when charging during the day and with our economic plug-in-field power, 255.82 kWh * 1.77 cents / kWh = <u>4.53 euros – for the whole month of car driving!</u> This means <u>great savings of 111.47 euros each month!</u>

Because it is so beautiful, here the power consumption price for 100 km driving with our e-car: 20.1 kWh * 1.77 cents / kWh = an extremely favorable 36 cents per 100 km!

The purchase price for the Nissan Leaf with 33,990 euros of course is still quite high. Nevertheless, the consumption costs are a very good argument for buying an e-car and when the purchase price goes down a bit more, the avalanche of electric cars, being the more economic and more modern vehicles, can finally break loose.

And, imagine, your "gas" station would then always be directly in front of your door and if the vehicles are equipped with an individual electric meter, theoretically it could be at each parking lot, whether at the supermarket or at your workplace.

The Result for Our Example Household

If we visualize that the example household – and all of us – have so far been virtually defenseless against any price increases for electricity, heating oil and automotive fuel, it makes me happy that this will no longer be true when our photovoltaic plug-in concept has been realized.

Here is the overview of the savings for the individual items and the total savings:

Expenditure Items Per Month (EUR)	Old Forms of Energy, Costs as of 2012	Energy Via New PV Plug-in System	Monthly Savings
Electricity	75.-	25.05	49.95 (66.6%)
Space heating	90.87	35.66	55.21 (60.8%)
Water heating	14.13	2.84	11.29 (79.9%)
Car	116.-	4.53	111.47 (96.1%)
Total result	**296.-**	**68.08**	**227.92 (77.0%)**

Overall, only 68.08 euros instead of 296.- euros per month would therefore be spent on energy costs! This amounts to a savings of 227.92 euros and 77%! For this amount, a lot of things could be bought each month.

Outlook for the Future Regarding Your Energy Expenditures

Now, you might say that the above is already an interesting result. If you look at the projected price increases of conventional forms of energy in the future, it gets even more interesting and a transition in the near future becomes financially even more attractive for us.

In the diagram of the example household we can see an average increase per year of the electricity price of 3.50% since the year 2000 (without the Renewable Energies Act apportionment). Regarding heating oil we had an average rise of 6.74% per year and with gasoline, the Germans experienced a constant price increase of 4.18% on average each year.

If we assume the same annual rates for the future, due to further shortages of energy resources and inflation, household expenditures for conventional forms of energy, logically, will cumulatively rise higher and higher to a surprisingly enormous amount in the future. In contrast to that, the installments for our plug-in-field power will always remain at the same low level.

Savings by the Three-Person Example Household Over the Next 30 Years

Let us have a look at how much money the household can save over the next 30 years. (All financial mathematicians: Please forgive me at this point that, for reasons of clarification we do not use the net present value method of calculation, but simply the sum of the annual amounts, which are thus not discounted again to the present value of the capital. Whereas, if we used the net present value method, the result would essentially be the same.)

Future Electricity Costs

For our calculation, we use the electricity price increase rate of 3.5%.

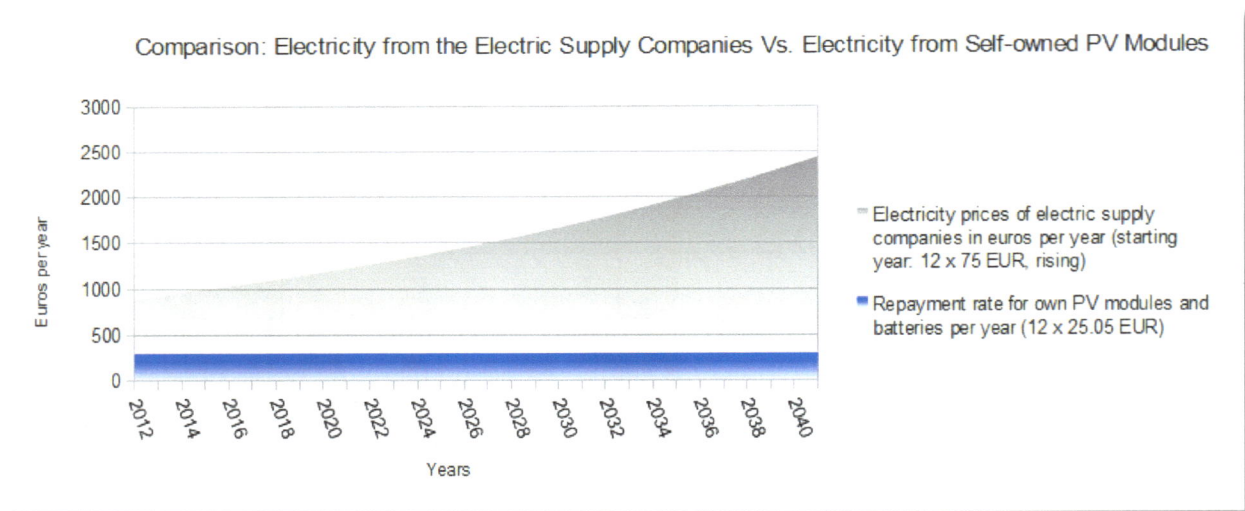

Comparison: Electricity from the Electric Supply Companies Vs. Electricity from Self-owned PV Modules

Electricity prices of electric supply companies in euros per year (starting year: 12 x 75 EUR, rising)

Repayment rate for own PV modules and batteries per year (12 x 25.05 EUR)

(Illustration: own diagram)

With the assumed rate of a price increase of 3.5% for conventional power, the accumulated costs for it amount to a total of 46,460 euros after 30 years, i.e. from the year 2012 to 2042. In comparison, self-owned PV modules and batteries, for the same period of time, only incur costs of 9,018 euros. This is a difference of 37,442 euros, which the example family saves!

Future Heating Costs

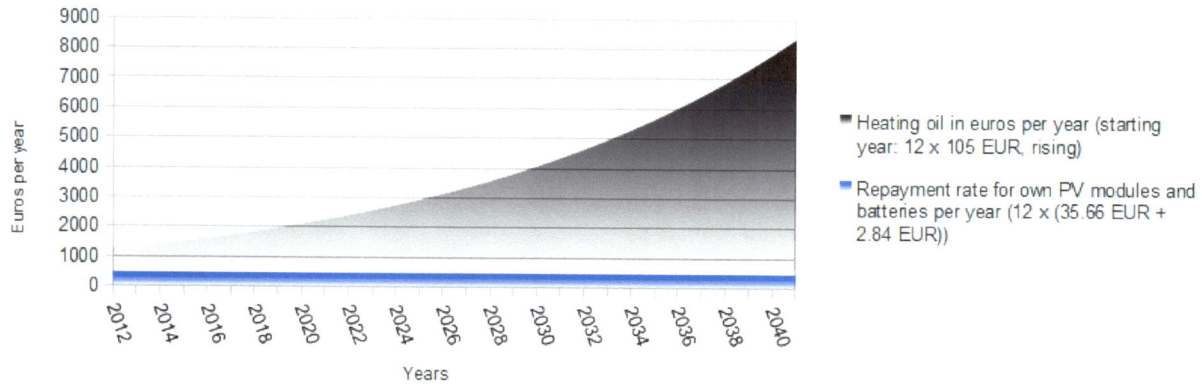

Comparison: Heating Oil Vs. Heat by Electricity from Self-owned PV Modules

Legend:
- Heating oil in euros per year (starting year: 12 x 105 EUR, rising)
- Repayment rate for own PV modules and batteries per year (12 x (35.66 EUR + 2.84 EUR))

(Illustration: own diagram)

At the assumed, relatively high annual rate of increase in the price of heating oil of 6.74% (as was the case each year on average over the last twelve years), the accumulated costs for this fuel after 30 years account for a total of 113,596 euros. For heat produced with self-owned PV modules only 13,860 euros in cost are incurred for the same amount of warmth. This results in a difference of 99,736 euros, which the example family saves!

Future Gasoline Costs

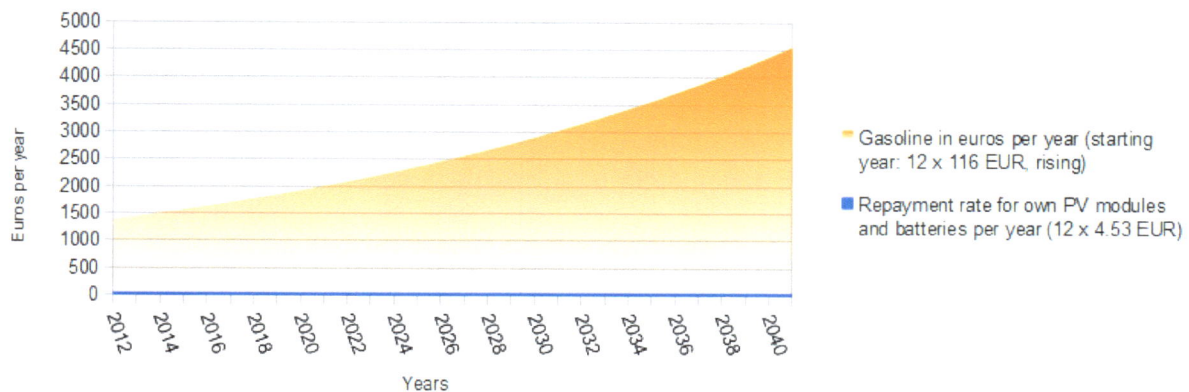

Comparison: Gasoline Vs. Electricity from PV Modules for the E-car

Legend:
- Gasoline in euros per year (starting year: 12 x 116 EUR, rising)
- Repayment rate for own PV modules and batteries per year (12 x 4.53 EUR)

(Illustration: own diagram)

With the assumed price increase rate for gasoline of 4.18%, the accumulated costs after 30 years amount to a total of 80,460 euros, while only 1,631 euros in costs are incurred for the the same amount of kilometers powered by self-owned PV modules. This results again in a big difference of 78,829 euros, which the family can save here!

Across all three items of expenditure, the family thus can save the difference between 240,516 euros (sum of the cost of energy from conventional energy forms) and 24,509 euros (sum of the costs of electricity from self-owned PV modules and batteries). As the result, we achieve savings of 216,007 euros or 89.81%! This amount truly can be described as considerable and we could buy many things with that!

Would you believe the result if you had not followed the calculation? Feel free to send me your answer on Facebook at the following link address: https://www.facebook.com/groups/photovoltaicpluginfields.international/

Conclusion on the Outlook for the Future

It is conceivable that the price increases for conventional forms of energy will be capped at some point in the future, because people would no longer accept such costs from the energy suppliers. Still, maybe this cap again will be cross-financed by taxes on other things, as is currently the case with coal and nuclear power, and as we mentioned briefly at the end of the first chapter.

Based on our three-person household example, we have now proven how much cheaper the photovoltaic plug-in concept is. The savings, which are to be achieved by this model speak for themselves. Furthermore, the photovoltaic plug-in concept is "powered" with completely clean electricity and in addition, the PV modules and batteries would be our own property that we could always sell again if we wanted to. Thus, there is much evidence that this concept can bring the solution to our energy challenges and to many environmental issues. So are you motivated to promote the introduction of this model as well? That would be great!

Still Further Advantages for You as a Participating User!

Would you have thought, that the rather "green-inclined" country Germany directly subsidizes the polluting coal and nuclear industry with 5.4 billion euros per year (Küchler/Meyer/Blanck 2012, p. 13)? How about your country? This money definitely could be invested in a better way – in you.

Further to the 5.4 billion, about another 35 billion euros in annual tax and insurance contributions could remain in the wallet of the citizens in Germany, an amount which at the moment must be continually spent to cover the so called "external society costs" of the conventional energies and only for this single country. The "external society costs" result from damage to the environment, to buildings, to the health of the people, etc., due to the many decades of burning coal and oil, and due to the use of nuclear energy. As a matter of fact, as long as the effects of oil, coal and nuclear still persist, the cost savings in this block will only develop slowly. But, the savings will begin to grow and desirably starting in the next few years rather than after several more decades. So we need to act soon and in every country.

Back to a direct cost savings effect that we have already considered in our calculations and is only repeated here because it is particularly interesting (to achieve a deeper understanding): One reason why we have so many fewer costs when consuming the self produced energy, is, because almost all parts of the electricity price that have nothing to do with the production of electricity itself do not need to be paid. These parts, which the price of the electric supply companies also contains, further to the production costs itself, usually include the following:

- Sales and distribution costs plus the energy companies' profit margins
- VAT
- Network usage charges
- Other taxes or apportionments

The number of items here varies from country to country, however, we can safely state that only a few of them would still be applicable with our plug-in system: The VAT would virtually be paid with the purchase of the PV modules. For the item "network usage charges" a partial charging would indeed be justified, but in our model these are supposed to already be fully covered by the fees of the 120 kWh / kW peak / year.

In addition to the financial advantages, we also have more practical advantages of the plug-in fields – here compared to the rooftop installation of solar systems:

- With the plug-in fields you have a service center at your disposal, in your county, with competent and helpful personnel for all questions relating to your modules, wind turbine shares and batteries but also for general questions about photovoltaic and wind power.
- You know that servicing, maintenance and snow clearing are always automatically taken care of.
- The interchange, respectively the selling of the excess power, is managed for you.
- If you move, you can simply take the modules with you to your new county, respectively this is organized for you.
- If you live in an historic, landmark building or in general do not have a roof available or it is not optimally inclined to the south, you can still, as already mentioned, supply yourself with as much of your own PV power as you want.
- You get 30% more yield by the solar tracking technique that is used on the plug-in fields.

With the plug-in fields and your own power plant there, you hence use the "oil well" in your back yard without the need to now fully cram your actual back yard with PV modules. By the way, feel free to already talk about this idea with your neighbors and friends and to discuss further proposals. On Facebook at the link address mentioned above, you can post them and get more involved that way. The more people talk about it, the sooner and faster the concept can be realized.

Benefits for Our District and Our Community

The positive effects on the communities and counties that the implementation of the plug-in-field system would bring are manifold.

As already mentioned, mounting racks or parts can be manufactured in the region. Or, if a photovoltaic or battery manufacturer is in the area and it offers the best value for money, all modules or batteries could potentially be bought there. Such things would directly support the local economy, more people would have jobs, and not least of all it would also increase the business tax income for the communities.

Further to the additional revenue on the one hand, the plug-in fields can also lead to a decrease in municipal spending on the other hand. This would be the case concerning specific social welfare expenses. Benefit payments could be adapted and would not need to be as high as in the past because, due to the introduction of the PV plug-in concept, the energy costs of welfare-recipient households would also decrease.

The county itself could employ people by hiring professionals for the management and the technical support of the PV fields.

Also, this workplace would certainly be an exciting field for volunteers.

The money, which the people would save through the new, very cheap self-produced energy – and we have seen how much that is – would more likely remain in the region. Up to now, it flows to energy companies and network operators that may even invest these funds abroad and not in your own country (some of these companies have their headquarters abroad, so it is logical that financial wealth of your country's citizens currently travels from your region out of your country and abroad). But if the money stays in the region, or in the community, a general revival will take place there. People will have more cash left to eat out at restaurants, to go to the movies, to visit cultural events, to play sports, to go to the public swimming pool, etc.

Through the aforementioned points, but mainly through the kWh-charges paid by the users, the plug-in field shall become a long-term source of income for the district. The county can profitably participate in the local interchange of power through the "kWh-revenue" that it gets from the participants and/or supply itself with (otherwise more expensive) power for all time. Alternatively, it can also resell the electricity to the county municipalities or at the electricity stock exchange. A short example calculation that you can adapt for your country: 200,000 inhabitants of the average county in Germany, times 7.25 kW peak on average per capita (the 7.25 kW peak per capita is obtained from the module amount of the three-person example household: 21.76 kW peak divided by three persons), times 120-60 kWh plug-in-field fees per year (60 kWh are forwarded for network use to the network operator and the municipalities) times 10 cents / kWh daily interchange price = 8,704,000 euros per year. Of course, the cost of the plug-in field itself and its management have yet to be subtracted. Nevertheless, it is expected to be a lucrative business, because the county will definitely have this revenue year after year. So in the end we have a win-win situation for citizens, the county and, as we shall see in detail below, for your entire country.

The plug-in fields, which would provide us with energy all year long, could also contribute to the cultural revival of the region. This could be achieved, for example, with events such as guided tours, information days or festivals regarding the topic of photovoltaic and wind energy fields. Due to such activities, a positive momentum often arises, in particular on a local basis, which one also can use again in return for the benefit of this meaningful infrastructure facility. A benefit could be to reach even more people for the project. Also, a "community effect" may thus actually develop. At the end of the day one forms something like a symbiosis and this in the, for us so important, field of energy supply (and certainly the people will then have a positive image of energy supply and of the suppliers as they are themselves the providers). In addition to electricity, the PV modules hence generate a sense of belonging as well, so to speak.

In general, clean energy production and its use for emission-free car transportation increases the quality of life in every region, whether in a big city or in a small town. To simply have the feeling, in our environmentally problematic time, to then live in a pollution free area, almost in a health oasis, would be a beautiful sensation. At the same time, and if your country belongs to the first movers, you can additionally say "my region is a high-tech region which the people of the world will certainly look at and be impressed."

Is There a Bitter Pill?

None which could outweigh the benefits of the plug-in fields. What could come to one's mind in this regard is the area required if each household opts for the cheaper, and in size slightly larger, thin-film modules (currently the cheapest technology: thin-film modules with amorphous silicon). However, what is already in sight and what will reduce land consumption, is that the efficiency per module will continue to improve in the coming years and thus less land will be necessary.

You are certainly asking yourself how we can estimate the necessary area at all. Here is the explanation.

For the calculation we use an existing thin-film module with the cheaper amorphous silicon technology. The relevant values are the rated power of 410 watts peak and the surface dimensions of 2.598 m to 2.198 m.

Let us use the area of Germany for the calculatory example. Suppose the Federal Republic of Germany consists entirely of three-person average households, i.e. households like in our example from the beginning of the chapter, then we would have 81.8 million inhabitants divided by three persons, is 27.27 million sample households, all having the same energy demand from the example.

Here again the energy needs of our average family and the resulting kW peak amount of PV modules:

Electricity for lighting, etc.	291.67 kWh per month divided by 30.5 days is 9.56 kWh per day
Space heating	415.12 kWh per month times 12 months (to get the annual amount) divided by 5 months (because we assume that we use the heating only during 5 months, i.e. November to March) divided by 30.5 days is 32.67 kWh for a winter day
Water heating	160.71 kWh per month divided by 30.5 days is 5.27 kWh per day
E-car	255.82 kWh per month divided by 30.5 days is 8.39 kWh per day

The total power consumption on a winter day therefore amounts to 55.88 kWh and on a summer day to 23.21 kWh.

You probably remember that we already used these numbers in the first chapter in the section "Strategy for the Winter" in an approximate manner. As a practical module amount we calculated 21.76 kW peak for the model family in that example.

Let us go on calculating the area required for the case of the 21.76 kW peak using cheap thin-film modules and for 27.27 million average families.

As an interim value, we determine the surface area of the modules per kW peak: 2.598 m * 2.198 m / 410 watts peak * 1000 = 13.93 m² per kW peak.

Before we can finally find out the total area necessary, we still need to take into account an important piece of information from the Fraunhofer Institute regarding the required space of PV modules if mounted on the ground: *"If facing south and with a sufficient distance in between them, they occupy about 2.5 times their own size."* (Wirth 2013, p. 35)

Now we have all the information we need and we can set up the final calculation.

Total area of all plug-in fields of the country:
21.76 kW peak * 13.93 m² * 2.5 * 27,266,667 average households = 20,662 km²

<u>In relation to the size of Germany of 357,120 square kilometers, the area requirement would therefore be 5.79%. The same percentage applies, of course also for the average county.</u>

Currently the efficiency ratio of the solar cells is constantly increasing. When thin film modules with twice the efficiency of the ones here used (efficiency ratio: 7.23%) can be made at the same price (in the laboratory better ratios are already achieved), the area requirement will be halved again. Then we will be at 5.79% / 2 = 2.89% of the area of an average county, respectively of Germany. For comparison's sake: Currently the share of agricultural used land in Germany is at 46.72% (Federal Statistical Office 2012).

Even if the area that is required for the photovoltaic fields perhaps is still large, there could be no better use for it! The two main reasons for this are obvious: We provide ourselves and others with cheap electricity and we can use it to heal our own environment! Slowly but surely!

By the way: Photovoltaic fields are also more CO_2-neutral than growing biofuel plants, because the fertilizing, transportation and burning of these plants produce more CO_2 than they absorb from the atmosphere during their growth.

The Benefits for Your Entire Country

Your entire country can benefit in many ways from the photovoltaic plug-in fields. On the financial, social and environmental level, the concept that we have in front of us brings us many solutions.

If the citizens are able to produce their own energy and are even able to sell it, the state for its part as well would increase its revenues by that. The people would have an extra income from which the tax authorities would in turn receive more income tax.

With our concept, not only would the entrepreneurial thinking of the population be generally stimulated, the economy would also instantly be boosted simply by the citizens acquiring the corresponding photovoltaic modules. Let's look at the result of the following rough estimate: If each household from our calculation above bought (financed) modules of 21.76 kW peak, we are talking of a purchase volume of: 27.27 million households times 21.76 kW peak, times 511.70 euros module price per kW peak, equals 303.64 billion euros. In addition to this we have the batteries, which would also be bought or financed. Regarding the battery capacity needed, we calculate with a modest amount if power for water heating and the electric car does not have to come from the battery in the evening, but if water is heated up during the day and the e-car is charged during the daytime as well with the PV modules. If, similarly, only a small part of the space heating comes from batteries, then on an average day, we have an electricity demand for the section of "lighting, cooking, appliances, etc." of 9.56 kWh (the amount is taken from the table of our example concerning the area calculation). We say again that we have 60% consumption during the day and 40% in the evening. So we need to store power for the evening equivalent to 9.56 kWh * 40%, which results in 3.83 kWh. For this we need at least two batteries from our example in Chapter 1, each with a 2.4 kWh re-delivery capacity and at a price of 1,309 euros apiece. Our volume for batteries is then, compiled for the whole country: 2 * 1,309 euros * 27.27 million households = 71.39 billion euros. The two amounts together, which would be invested by the citizens add up to 375.03 billion euros. About one-third of the sum can be further added due to the investments in wind turbines (depending on the demand), and a tenth of the 375.03 billion euros can be calculated on top because of the infrastructure to be built by the counties. <u>Overall, only for this one country, a volume of roughly more than half a trillion euros could be expected.</u> These investments, in preferably national products, would certainly have a strong positive impact on your economy and the number of employees in your country would most likely increase due to this concept.

In addition to the new income tax revenue, from the achieved economic growth and from the steady electricity selling of the plug-in-field participants, the state would also receive a significant revenue boost from the value-added tax, which arises when the modules and batteries are purchased and shares of wind turbines are acquired. Many billions of euros would be earned by the state in a very short time. Such an "excess" of income could ideally be used to reduce national debts and in consequence to reduce the debt-related interest payments. In this way the financial situation of any state can be eased.

The tackling of this project would be a great image boost for each government and each parliament in the perception of their voters because the politicians would prove great strength and energy and would demonstrate that they have the will to bring together all people in the country: people who can invest a lot and people with little money, tenants, homeowners, business owners and workers. The government would combine economy and sustainability in the best manner. An example that shows that the people are interested exactly in that, is the rapid expansion of solar systems on the roofs of their houses. In Germany, as of 2012, 3.08 million solar power and solar thermal systems have been installed (German Solar Industry Association 2013). Another example is the development of energy cooperatives. There are now more than 600 in Germany (Renewable Energies Agency 2012b).

The export of produced electricity could also be very beneficial for your country because, in this way, a lot of additional money from abroad would flow into your voters purses and into your economy. It is obvious that it would be a huge business if all your country's citizens exported electricity. International electricity trade does have the potential to bring nations closer together as well. Each of your neighbor countries should have an interest in buying clean electricity from you at a fair price, for the benefit of their own citizens and instead of burning coal and oil or producing nuclear waste. When sun or wind is available, the electricity from these sources is cheaper than the electricity produced in conventional coal or nuclear power plants. So it is, economically speaking, logical that they will buy from you. And, because you would help your neighboring countries to save CO_2 and nuclear waste, it is therefore to assume that the positive image of your nation would be further enhanced on your continent and in the world.

What certainly will please the minister of finance and the budget committee of your parliament in particular, beyond the previous points, is that almost no financial support is needed by the government, except for the low interest rates of 0.5-1% for the loans of your national development bank. So the concept is very budget friendly on the expenditure side. Moreover, there is virtually no creditor risk, because in awarding such loans, the government invests the tax money back in the taxpayers. Both government and taxpayers can therefore agree in saying: There is no better and no safer use for loans than in the allocation with its own citizens!

Another pleasant effect for citizens and the government is that the people can actually work less and keep the same living standard as before. This would be possible because of the several hundred euros they save on energy costs each month and by the simultaneous earnings from the electricity sales. The weekly working time could therefore be somewhat reduced, if desired. The people would have more time for themselves and their families or for necessary professional training. If then the employees were to work fewer hours per week, more people could be put to work. The unemployment agencies would therefore save millions in the payment of unemployment benefits.

For countries that are paying feed-in tariffs for renewable energy production, a declared aim of their policy often is to stop or reduce the payments again as feed-in tariffs cause additional apportionment costs for the consumers on the other end. Our "self-consumption" concept makes the instant achieving of this goal possible because the system is already profitable by itself and so it can work without the feed-in tariffs.

If people in the future produce their own electricity, the question of who will pay the feed-in tariff apportionment will arise in the nations with existing long term feed-in contracts (in the case that it is financed via the household electricity bill). Three proposals to compensate the apportionment, that could be used in combination as well, are the following: For buyers at the electricity stock exchange an electricity market price surcharge could be introduced for the cheap solar power. So if PV electricity is traded for less than, say 5.3 cents / kWh (average market price of electricity at the leading European electricity exchanges from 01/2008-10/2012) (Küchler/Litz 2013, p. 4), the buyer pays the difference up to this value into the feed-in tariff apportionment account. In this way, the industry sector is engaged as well but no one is discriminated against (in Germany the apportionment mostly is to be paid by private households, the industry is usually exempted from it). For households that will not have migrated to self-produced electricity, thus the previous apportionment would then only need to rise slightly or not at all. As a second proposal, also a part of the income tax from the sale of electricity could flow into the apportionment account. Furthermore, there is the opportunity to create incentives so that today's solar system owners, who receive a feed-in tariff, will become consumers of the electricity they produce and so for the future waive the receipt of the feed-in tariff. This would also lead to a strong shrinking of the apportionment. We will return to this point in the next chapter.

With the decision to implement the plug-in concept, your country would opt as well for a better allocation of resources than if, for example, investing in the famous Desertec project. In my opinion this is true in the same way for each company that is involved in Desertec or is considering getting involved. This is not to say that investing in Desertec is not useful, but it must be said that the investment of resources and manpower in the plug-in project is more useful and more advantageous. A brief explanation of Desertec: The Desertec project is planned to emerge from a number of individual projects of solar and wind power plants in North Africa and in the Middle East. With investments of around 400 billion euros approximately 15-20% of Europe's electricity demand is to be covered by the project by the year 2050. Eighty percent of the electricity is to be sold in Africa. Supposedly a laudable idea, but there are still several decades remaining until 2050. It is also not clear yet at what price the power will then be sold in Africa and if the people there will be able to afford this electricity. It is also planned to invest mainly in solar thermal technology, which has already been overtaken by the photovoltaic technology regarding its price-performance ratio. And from Europe's point of view, 15-20% coverage of the electricity demand is not very much compared to the time frame that remains until 2050. In addition, the political situation in the countries, in which are being invested, is not always stable and the safety of the plants is therefore not necessarily guaranteed. With our plug-in system, however, the power consumption of the whole of Europe, Africa, Asia, Australia, South and North America can already be covered in a few years. Estimated, after a phase of about two years of planning and testing, in about another three years. Consequently, the companies involved in Desertec could make an earlier and more reliable profit if they would invest their partial amounts of the 400 billion not in Desertec but instead in the plug-in-field concept, e.g. in increasing the efficiency of the photovoltaic panels and batteries. In a further step, the more efficient technology could then be sold in Africa and in the Middle East and as well to the private individuals there. In these regions, the plug-in concept could just as easily be used and unlike with the Desertec philosophy, which is based on large centralized power plants, the savings and revenues would surely reach the common people.

As for the blackout resilience in your country, with the introduction of the plug-in principle a widespread power failure is virtually ruled out as for every citizen the electricity is produced in the immediate vicinity, so to speak. No one can cut off your country from the power for even a day because every day the sun rises and plenty of light and wind for new energy reaches you again.

There would also be no energy dependence on other countries anymore. This is especially beneficial if your country currently buys a lot of energy commodities from politically unstable regions, or from regions that do not have the human rights standards which exist in and are expected in your country. In case your country's rate of import of energy resources is not known to you at the moment, you can certainly find it by doing research on the Internet. You might be surprised how high it is. This dependency and the outflow of money can belong to the past once every household has installed its photovoltaic panels in the plug-in fields. Should then, especially in the sunny season, an excessive photovoltaic power production occur, then, via the Power-to-Gas storage, the gas reserves can be increased or energy can be exported in a steady manner. This means that if "too much" power were produced, it would virtually never be a problem. The power does not need to be given away for free anymore, as is sometimes currently the case. In this way your country can turn from a previously dependent energy importer, with a possibly very high import share, to an energy exporter with profits for citizens, businesses and the state, and with a secured supply perspective for all time.

Apart from exporting abroad, the private households could supply the complete industry sector, services sector, trade and transport sectors of your country as well. Of course in this case, the households would need to extend their PV module and wind power equipment correspondingly. Naturally, the businesses themselves can also set up their own photovoltaic modules. Once the companies realize that the plug-in system works smoothly, they will most likely, for reasons of profitability, also jump on the bandwagon.

The advantages of a country that uses only clean energy, for electricity supply and heating as well as for car traffic, are manifold, not only on the financial level and for health reasons, but also, and this is no less important, on an emotional level. Our sense with which we feel the quality of life most probably tells you that the quality of life would increase greatly. Your country would then be a place where you can feel even more comfortable, healthy and safe. We would have a new, positive perception of our environment as well as of the people around us and a new, positive attitude about life would be given momentum. It would also pacify, at least partially, the environmental conscience that we have for today's world and for the world of our children. Many of us have this conscience together with a very strong sense of responsibility, as we do not only want that we are doing well, but also that everybody else is doing well. So far, there have been only a few opportunities to act on because we have been practically dependent on the harmful oil, coal and nuclear industries for our daily supply of energy. With the concept here front of us we can finally produce all the energy we need, every day and in harmony with nature. A reassuring and beautiful feeling.

Here, once again, for reasons of completeness, a brief overview of the advantages for your country already mentioned in previous chapters:

- Nuclear power and its risks would more quickly become completely unnecessary. Also gradually in neighboring countries.
- The unnatural CO_2 emissions, except for air and sea transport, could be reduced to almost zero. We would have virtually no CO_2 emissions anymore in the following three areas:
  - Electricity supply: The goal is to reach 0% CO_2 by 100% renewable electricity from an efficient domestic production, mainly by photovoltaic and wind power, but also still by hydro power, biogas and geothermal energy.
  - Heat supply: When the electric power becomes cheaper, due to our system for photovoltaic energy, heating with electricity becomes more meaningful and more cost-effective. Expensive central heating systems and rising oil prices would be a thing of the past, in favor of infrared heaters, which then would use completely CO_2-free electricity.
  - Transportation: With the avalanche of PV electricity, the avalanche of e-cars can finally arrive. The e-cars will then simply be the more modern and less expensive choice. Electric cars like the Nissan Leaf already cost only 5.49 euros in consumption per 100 km (consumption: 20.1 kWh per 100 km at 27.3 cents / kWh [Houben 2012, p. 2]) and they can function completely without CO_2 emissions in daily operation. By the way, electric mobility has already begun in a widespread manner on two wheels: Until now more than 1 million e-bikes have already been sold in Germany. But also about 50,000 hybrid cars and several thousand purely electric cars are cruising on German roads to date (Wirth 2013, p. 56). As for the heavy goods vehicle traffic and because of our extremely low PV electricity price, the truckloads could increasingly be transported by rail – as well for the benefit of freer motorways!
- A positive impact of our plug-in PV fields, which can not be mentioned often enough, is: finally being able to breathe good and healthy air in our cities! Is this a feeling you do not know either? How about if the traffic fumes and smoking chimneys would be gone within the next few years? Sounds good, doesn't it? Cities would thus become oases of fresh air, just like almost any of the popular climatic spas in your country!

All aspects together would mean that with the application of the plug-in fields, your country would be among the most advanced nations of the world, with even greater prosperity and a higher standard of living than before. Maybe you would even become the world leader in the export of this idea and the associated technology. Since this system is quick and easy to use by almost all nations of the earth (whether they have a comparably high level of solar radiation or not), it pays off when you are the fastest on the market and the first country to gain practical experience with this concept. This way you can offer proven products and systems together with the best know-how and deliver directly.

The Benefits That Our Concept Would Bring to the Whole World

There really is a benefit to the whole world. Decide for yourself how great it can be.

A great advantage of the county plug-in system is that it is applicable anywhere in the world. How exactly the administrative districts taking care of the plug-in fields are called in the different countries, is irrelevant. Important is though, that the principle works in urban and densely populated areas as well as in minimally populated areas, which can even lie detached from any national power grid. So, the people will be able to set up an energy supply in any place, which allows them to build all kinds of industries in their region. And, it is only with electric power that businesses can produce in an efficient way and so compete in the twenty-first century business environment – whether now the very latest machine equipment is available, or whether it is "second hand" – without electricity, people inevitably remain, so to speak, in the eighteenth century.

With our model an agricultural boom could be initiated as well in areas that would normally be too hot and therefore not fertile enough. Through plug-in-field-powered desalination and irrigation, new agricultural oases and granaries can emerge in regions previously hallmarked by malnutrition.

If people, in areas that were formerly depleted or had little wealth, now have a small but growing economic basis, and more wealth is promised, then they have something, which they want to continue to build, maintain and protect. These are real roots needed for the stabilization of regions that were previously unstable. Despair and existential fears disappear, in favor of hope and the confidence of a secure and comfortable future. People will start to care more for education and more economic growth will follow. An educated middle class can arise, which has interest in a stable democracy and a fair rule of law in order to protect the earned goods and freedoms. From the perspective of any open minded nation, this means that there would be more partners in the world with interesting markets to buy from and sell to.

In general, one can also conclude that there will be less emigration. Many areas that are still poor to this day and which unfortunately could not provide sufficient prospects to the local residents, will then have, with the photovoltaic plug-in fields, a basis on which to build further.

A most important achievement for the world would be that wars for energy can belong to the past because each state will produce enough energy by their own households! Our electricity concept, which is intended to create cheap and clean electricity in abundance, can so also bring peace and stability to many regions. This "by-product" would therefore be an even greater gift than the actual and original purpose of the model.

Chapter 3

The Steps to Implement the Plug-in Concept

Right now, the concept described in this book is just an idea but it is an idea with which we can reduce the price of electricity to a third of its current price. Already therefore, the implementation is more than beneficial for any nation. The ideal way would be if the concept was introduced in the parliament of your country as a national infrastructure project with the application for implementation.

A big project like this needs to be thought through, discussed and tested very well. All public bodies which are affected in some way by this new energy system must be heard. Just as the citizens are to be heard.

This project should be as transparent as possible. With an open dialogue between policy makers and citizens, on the basis of this enterprise, the opportunity to strengthen the image of all parties sustainably in the eyes of their voters comes as well. People want to be taken on board and they want to participate in the economy!

When this idea has spread and has finally reached your parliament, the further steps to implementation could be as follows:

1. Discussion and allocation of competences in the parliament
2. Beginning of the planning in a parliamentary committee, choosing of an expert commission by the committee and providing of the frame conditions
3. Detailed planning of the implementation by the commission of experts, planning of the advertising and of the continuous information about the project, selection of the field-test counties and the conducting of the tests
4. Discussion of the results in the committee and in the parliamentary plenum plus voting concerning open decisions in the parliament
5. Implementation on a large scale
6. Constant follow-up and adjustments if necessary

1. Discussion and Allocation of Competences in the Parliament

The photovoltaic plug-in concept is introduced in parliament as a proposal for the nationwide launch. This can be done by one party or by members of parliament across party boundaries. After that, an open discussion in the plenum follows and a committee is determined, which is responsible for the efficient implementation.

2. Beginning of the Planning in a Parliamentary Committee

Regarding the question which committee might be the most appropriate for the mission of planning and implementation – taking into account the thematic competence, this would be the committee of the department of energy and/or economics and/or technology.

Nevertheless, our energy concept touches the areas of competence of various ministries, all of which would have to be heard in the committee responsible. These ministries are of course the ministries of energy, economics, technology, environment protection, building and agriculture.

Equally, the competence of various public authorities would be required, so it would be important to also listen to their experts. Among others, the agencies responsible for the national energy supply and the national electricity network are meant here. Communal associations would also have to be involved, such as the county association and the association of cities.

Choosing of an Expert Commission

After this round of consultations, a broad expert commission would be selected for the detailed planning of the large-scale project and to carry out the important field tests. The commission would scientifically determine, among other things, which would be the most efficient technical systems, what would the most practical way of organizing later for the counties be, how could citizens be involved, informed and supported in the best way, etc.

Here are some suggestions from which fields and institutions the individual experts could come from, respectively of what kind of renowned professionals the commission could consist of: project planners from the solar industry, manufacturers of photovoltaic modules, mounting frames, control systems and batteries, members of the counties and the cities associations, municipal energy suppliers, energy cooperatives, network operators, scientific institutes, university professors from the fields of energy, economics, business administration and marketing, etc.

Employees and agents of conventional energy companies would have to be heard but their opinion would have to be interpreted with caution because of the conflicting interests due to the loss of their oligopoly due to this concept.

Providing of the Frame Conditions

The committee would also take care of creating the frame conditions that would be necessary for the implementation. There are various political and legal issues that would have to be addressed:

● Making legally sure that the electricity, which would be produced, consumed and interchanged by the private households, would be disburdened from taxes and apportionments that are not applicable anymore.

● In chapter 2 we already discussed the question of who would pay the feed-in tariff apportionment (for the current subsidizing of the renewable energies – in case your country applies this measure) if the households produced their own power to an extent of up to 90 or 100%. To cover these apportionment costs, there would be a number of possible means.

One method would be the already mentioned electricity market price surcharge for the cheap photovoltaic electricity (up to the amount of the average market price of electricity), which would be payable by the purchasing company.

As also already discussed, a part of the income tax of the profits of the electricity interchange and sale by the households could be used for the payment of the apportionment.

Another tool is that current recipients of feed-in tariff payments could be offered incentives, so that they would waive the contractually guaranteed future feed-in payment amounts. How could this work fairly for all parties? The "switchover bonus" in the proposal outlined here would most purposefully be paid in PV modules and batteries: The "feeder", who switches to become a consumer of the self-produced electricity, would receive so many PV modules, as a gift, that he could earn as much money as before in terms of kilowatt hours saved on his electricity bill.

Example: Assume that the current earning of the feeder is 500 euros per year, then he would get so many PV modules (for the installation on the plug-in fields) that they would save him this 500 euros annually at the current price of household electricity. We calculate: 500 euros divided by 27.3 cents / kWh (household electricity price at the beginning of 2013 in our example country Germany), equals: 1,831.5 kWh. So he would receive as many modules as necessary to enable him to produce 1,831.5 kWh per year. For Northern Germany, that would mean 1,831.5 kWh / [900 kWh "possible electricity production quantity in northern Germany per kW peak" + 30% more yield by solar tracking - 187.85 kWh plug-in-field fees (the plug-in-field fees were calculated in the second step from 120 kWh / kW peak charges)] = a module set of 1.86 kW peak as a switchover bonus. Expressed in euros, these would be PV modules with a value of 954.21 euros. As an added incentive, everybody who opted for the switchover would receive the necessary batteries as well so that they could use the generated power at night also. Here a battery with at least 2 kWh re-delivery capacity would be necessary (calculated for 40% evening and night consumption), i.e. 1,309 euros in battery costs (as in our example in the first chapter). To repeat: With the above the "switcher" would get 1,831.5 kWh of electricity per year, available day and partly at night, for the service life of the modules and batteries! In addition, he would obtain the power generated by his "old" modules, which could easily amount to several thousand kWh as well. For the party of the apportionment-payers this would be much cheaper than paying 500 euros each year for up to 20 years, as would be the case with this example. And, for the party of the previous feeders it would make no difference in terms of the return. To avoid feeders postponing the decision to switch too far into the future, for each year that the feed-in contract of the respective individual had already been in effect, one twentieth of the switchover bonus (depending on the duration of the contract) would be deducted.

Another source of money for the feed-in tariff apportionment funding could be tapped by making the energy companies contribute to the costs of nuclear waste and CO_2 emissions in a more relevant manner, rather than having the ordinary citizens bear these costs through their taxes. This way the taxpayers would have still the chance to get back something of their "deviated" tax money. The released tax funds could so be used for the apportionment in case that the previously mentioned measures were not already doing the complete job.

- To temporarily boost the plug-in concept in the population and to not relatively discriminate against pioneering buyers, they should be rewarded for their role of leading the way. And this would be by allowing them to pay today only the amount for their PV modules, which they would merely pay after two more years of further technological development. In other words, they would buy today but pay the lower price of two years in the future. In this way, a critical mass of pioneers would rather buy today instead of waiting for the price to decrease some time in the future. In the first year, for example, two modules could be given for free for every ten modules purchased (free products are generally more popular with customers and more haptic than mere price reductions). This promotion incentive could be funded by one of the measures described above for the apportionment pay back, respectively, the incentive would even be free for the state, as the state would virtually only have to waive the 19 percent VAT for the modules and for a limited period of time.

- In terms of international frame conditions, the committee should also clarify whether new international agreements with the neighboring countries are necessary. Especially for the interim storage of power in the neighboring pumped storage power plants. The committee could even negotiate discounts, because of the large capacity that perhaps could be drawn on.

3. Detailed Planning of the Implementation by the Commission of Experts

The detailed planning would be done by the expert commission. When it were completed, it would again be presented to the parliamentary committee for discussion.

<u>This book is intended to serve as the concept of the detailed planning. It is recommended as a "blueprint" of the infrastructure and of the processes of this ambitious undertaking.</u>

Of course, in case the project is implemented, there will be many additional and optimizing ideas before, during and after the field tests by many people. The expert commission, together with the committee of the parliament, will then ideally accept any proposal for evaluation.

Planning of the Advertising and of the Continuous Information About the Project

The two main goals of the communication are first, that citizens know about the project and are conscious of how it works, and second, that they adopt the project on the broadest possible basis and participate in it. The communication plan, which the expert commission would establish, would therefore need to enlighten the people and educate them in relation to the project. It would always have to inform the people of the current implementation status and on the developments, and it would have to inspire, so that the people would become more and more interested and motivated.

Hence, ideally, there will be two permanent campaigns, which merge into each other and support each other: an information campaign and an advertising campaign.

The advertising and the information campaigns must be present in all common media, i.e. television, radio, print and Internet. Especially on the Internet, the participant shall find all the information that might interest him or her in a well structured manner.

It is important that the citizens are involved early, before the nationwide launch, so that the campaign can be successful. For example, the carrying-out of the field tests will be filmed as a television documentary and be shown as multiple episodes in various science programs. In this way, the citizens can be made closely familiar with the plug-in-field system before its national introduction.

The financial resources for the advertising campaigns can come from either the government or from the interest groups, such as the solar industry associations or the associations of energy storage and many others.

If we now think about what the advertising message can look like, help comes from an unexpected direction: It is the high price of the conventional energy forms, which provides the best argument for citizens to participate in the plug-in fields. Who would have thought that the expensive current household electricity prices and the high cost of gasoline and heating oil have something positive? However, they make our solution more attractive and thus unintentionally contribute to a faster changeover of the citizens towards their own plug-in PV power.

What else is important when the advertising message is being developed? There are many other benefits that may be included in the communication, e.g. that the modules and wind power shares supply enough energy for each household; that they do this every day, and for many decades; that the supply is as secure as light will come again the next day because the sun always rises; that the county government will take care of everything; that there is a community of power interchangers; that the plug-in fields are the reason why PV power, as of now, is so economic; that the participants can reduce their energy bill to the desired amount, thanks to the financing; that a lot of money remains in the pockets of the people so that they can buy many other things instead; that it is the state, which makes all this possible for its citizens and that a new, positive feeling and world-view and a more satisfied environmental conscience can find its way into our lives again, etc.

People can receive all this and very easily. They just need to go to their service center at the county administration for a consultation. There, they can order the photovoltaic modules and, if they want to, they can be there when the modules are plugged-in on the plug-in field. From that moment on, they have their own power station for their consumption, their heat demand and their electric car. In this way, they will save themselves approximately two-thirds of all their energy costs – and this regardless of whether they are tenants or homeowners. Moreover, they can be proud to be an important pioneer and participant in the largest clean energy project of humanity.

So these are already many advantages that can go into the advertising text of any radio or television spot and could be phrased in the following ways.

Example 1:

The government now helps you to reduce your electricity price to a third with the largest clean energy project in the world!

Provide yourself, as of now, with as much energy as you want – for your home, your heating and your electric car – every day for all time – and with the utmost security of continuous supply, guaranteed by the sun with its daily light – and managed by your local county government.

Finance your photovoltaic panels, batteries and wind turbine shares for your own 9-cent-per-kWh electricity now! The financing of the equipment is offered at a low 0.5% interest rate – guaranteed by the state!

So just have your modules and batteries plugged-in at the new county plug-in fields and benefit from your own power station or interchange the generated power profitably in your county community.

The fees, which are charged in kWh, are only 120 kWh per installed kW and per year. Advantageously, the fees are earned back automatically up to three times again through the solar tracking technology, which the plug-in fields are equipped with.

So, thanks to the infrastructure provided by your county, you have a higher yield than with a system without solar tracking. And yet, you have much lower costs, because you merely buy the bare modules and batteries, which you purchase at the very economic manufacturer price – guaranteed. Of course, the installation, servicing and maintenance of the equipment are also taken care of for you.

Thus, you will save two thirds on your conventional and growing household electricity price of currently on average 27.3 cents per kWh! That is to say, just determine yourself, through small installments, how low your monthly energy costs will be. In this way, they remain stable and favorable, in the future and for all time.

And, if you decide to participate until the "DD of MM of YYYY", you will receive 2 photovoltaic modules extra for free for each purchase of a set of 10 modules. With this bonus we would like to thank you for being a pioneer and because you are leading the way, together with us, into a new economic and clean energy future!

Call now or visit the nationwide information website at www...

Example 2:

Save two-thirds of your electricity price as of now and fix the price at this low level for up to 30 years! How it works? Simply finance your own photovoltaic modules, batteries and wind turbine shares, which you can purchase from your county, at the very economic manufacturer price. This is where your modules and batteries will be plugged-in later as well. [Very important: It should sound as easy as it actually is.]

Your county provides the complete infrastructure for a fee of only 120 kWh per kW of installed capacity and year. Since the 120 kWh fees will be recouped again by about three times due to the county's solar tracking technology, using the county's infrastructure, so to speak, results in being free.

Together with your night storage batteries, you then have your own cheap electricity available for you, around the clock, on an average of about 9 cents per kWh.

Just contact your local county service center. The county staff will be happy to answer all your questions regarding photovoltaic modules, night storage batteries and wind turbine shares.

Another big advantage: It does not matter whether you are homeowner or tenant, the power flows to your electricity meter in your home.

As already mentioned, you can purchase and finance the photovoltaic panels and the night storage batteries at your service center, at the very economic manufacturer price (!). For the financing, your government offers you a loan at a favorable 0.5% interest rate. How do you like that? Are you interested?

And the best part: In the pioneer year, when purchasing 10 photovoltaic modules, you get 2 modules for free on top – but only until the "DD of MM of YYYY"!

So, reduce your energy costs from now on by two-thirds and become independent from all future electricity price increases! Households that still buy conventional electricity, currently pay 27.3 cents per kWh on average. With your own photovoltaic modules, the cost per kWh is less than 9 cents on average. So simply get your own energy!

By the way: Since you consume your own power, you do not need to pay the feed-in tariff apportionment and many other taxes and levies.

So call now or visit the nationwide information website at www...

These were two examples of possible lines of argument in order to imagine how it might sound afterward in the media. When we then turn on the radio or the TV and can hear or see advertising for our beneficial energy project, we can rightly be proud of ourselves and our country and know that we did it and implemented this beautiful enterprise.

Selection of the Field-test Counties

The counties themselves shall be able to apply to become test counties for the trial of the plug-in system. An application by the county itself indicates a great openness of the participating municipalities towards renewable energies – which of course is desired. As soon as there is a narrowed down shortlist of applying counties, their citizens shall be interviewed in a representative manner as to how "pre-interested," they are as well.

Then the ideal two counties will be selected from the shortlist and the field tests initiated by the experts.

Conducting of the Tests

The field tests are of enormous importance to gain valuable experience before the big launch. In particular, they will provide information on the best way of organizing the processes on site, the most efficient technologies to use and furthermore, on the best form of communicating with the participants.

Some proposals on the frame conditions of the tests:

- For the field test, it is recommended that the expert group be split and thus two teams be formed, both covering the same fields of expertise. Each team will then take care of one of the two test counties. It is important that both teams share their experiences during the test continuously. By using two teams for the study, there is the advantageous effect of getting two different perspectives on the project. Something like a "sportsmanlike competition" might even arise between the two groups, which can inspire the developers' spirit and lead to an even more meaningful overall test result.

- The test counties should be clearly addressed and assured that the system of plug-in fields, which is implemented for the test, will definitely remain in the future (contractually declared to every participating citizen – as it should also be later for the nationwide implementation). This is important so that the residents do not have to worry asking themselves: *Will the plug-in field be looked after as well in the far future or just for a limited trial phase? What kind of guarantee do I have if I invest my money already during the test phase in it? Etc.*

- While the infrastructure for the test is created, the citizens need to be constantly informed about the progress of the project through the local media and at the same time they should always be invited to make proposals and to give feedback on how they feel about the implementation. The proposals and the feedback will be constantly evaluated by the team of experts.

- It should also be said to the people that the participation in the plug-in-field-powered system is not obligatory but it is an offer.

- It is also very important for the test that the photovoltaic modules be as cheap for the participating "test citizens" as they will be on the large-scale future implementation date. Therefore, a little subsidy will be given to the test participants, being so to speak "pre-pioneers", in order for them to not have a disadvantage compared to the later investing rest of the country. So, assuming that the nationwide launch date will be exactly one year after the test phase, a discount of 10-20% can be expected during that time, which will be reimbursed to the test participants directly after the purchase of the equipment.

- As already mentioned, the implementation of the field tests shall be broadcast as TV documentaries. In this way, the citizens can become more familiar with the plug-in-field system before its national introduction.

At this point, a few interesting elements, which special attention should be paid to during the test phase. They should be examined by the expert team on practicality. The experts can and should make adjustments during the test.

- Instead of using small battery solutions, it could be analyzed in the test, whether large-scale storage units are more efficient, even if they require a more complex cooling mechanism. The large-scale storage units could then be acquired by the participants in the form of shares at the county. If the participant moved away, he would get the residual value of the shares paid back.

- The amount of modules, which a household wants to install shall not be limited. The service center of the county advises the participants on how many modules they need for the desired amount of self-produced electricity. In the test, special attention should be paid to the issue of how many modules the households want to buy beyond their personal needs (for the purpose of generating an additional income by the exchange or sale of their electricity). This indicator is important for the estimate of the total area required, later at the national level.

- As an essential result of the test a mounting rack system must be found or developed on which different PV modules can be easily mounted. In general, the finding and determining of norms is useful and facilitates the subsequent mass production of all components.

- Once the modules are mounted, each participant will have a small digital electricity meter attached to his set of modules. This meter continuously measures the electricity generated and at the same time transmits the values to the Internet for monitoring. Depending on the price range it might even be efficient (which would have to be investigated) to install a small electricity meter on each individual module. This would allow a quick identification in case one panel had a technical problem at some point in the future.

- The flexibility of the whole system should also be examined carefully. The participants should always be able to buy more modules later and to have them added to the plug-in field. There could be a minimum number for each request to add new modules, for example from five modules or more. Some households might have only recently purchased a conventional heating system or a gasoline car and may not want, or be able, to replace both right away.

- The consulting in the service center is elementary and so is therefore, the training of the county employees. The field tests will help in finding the required contents of the training so that the service center staff is able to optimally counsel the prospective participants and to determine their needs of photovoltaic modules and batteries.

For the assessment of the needs of the prospects, the employees ask for the current power consumption habits. Here a pre-selection of questions: *How much electricity consumption did you have last year? What is your ratio of day and night consumption? Do you already use energy saving light bulbs and appliances? What type of heating do you have installed?* (when people are interested in switching to infrared heating and/or to an electric car, the service center can also give an initial consultation regarding these topics) *How much are your current heating costs? How large is your living space to be heated? Is there still potential for thermal insulation or do you have a well-insulated house? What car do you drive? How old is it? What are your gasoline costs? Do you drive mostly short distances? Etc.* (More questions will be selected by the expert team.)

The responses are then inserted by the service center staff into the calculation tool on the central website. In this way, the required amount of PV modules, batteries and wind turbine shares is computed. At the same time the website tool itself, which is publicly available, is explained to the client, so that the client himself can find it, use it, and later also explain it to other interested parties, at any time. For this part a certain pedagogical training session would then probably be necessary for the employees.

The above and other points should be investigated for a perfect country-wide implementation.

After the trials are completed, the expert commission draws its conclusions from the test results and based on these it composes a strategy paper for consultation in the parliamentary committee.

4. Discussion of the Results in the Committee and in the Parliamentary Plenum Plus Voting Concerning Open Decisions in the Parliament

The test results, the conclusions drawn and the written strategy paper will now be discussed in the parliamentary committee and, where appropriate, the strategy paper will be adjusted.

After the subsequent presentation in the parliamentary plenum and after the necessary voting in all the required decision-making bodies, the final and binding version of the strategy paper can be formulated.

Based on the results of the field tests and based on the final strategy paper, an implementation guide for the counties will then be developed. This guide will act as a manual and should answer in a simple way, all questions regarding the optimal realization of the plug-in fields. Of course there will also be intensive trainings for the counties as well as a hotline to a nationwide team of consultants, who will help to answer all remaining questions that the county staff might have.

The preparations for the nationwide launch will also be made on each involved political and public authority level and as well on the basis of the final strategy paper.

Regarding the distribution of the operational authority, it would make sense that the national energy or network agency should have the technical responsibility and the county association should have the responsibility for the administrative questions. Both authorities would have to work closely together.

5. Implementation on a Large Scale

It is recommended that at least two months before the start, each county has created the necessary infrastructure and has tested it for proper functioning. An official technical inspection authority or the national energy or network agency must confirm the functioning and the fulfillment of the security requirements in each case.

To balance the electrical loading of the national power grid, it is suggested that the countrywide introduction of the system is done successively, e.g. thirty counties at a time and one week apart.

To avoid long queues of interested people in front of the service centers of the counties, the counseling interviews ideally are to be arranged in advance by telephone.

Since there might be a big rush when it comes to ordering the modules and batteries, delivery times of several weeks naturally may occur.

6. Constant Follow-up and Adjustments If Necessary

During the launch and beyond, there should be constant follow-up checks and, if necessary, adjustments to the infrastructure, to the organization and to the communication. Such adaptation needs and suggestions of all kinds are to be collected centrally and discussed by a board consisting of all parliamentary committee members, representatives of the national energy or network agency and representatives of the county association.

Now, in the way it is described in this chapter, we should be able to accomplish the implementation of the plug-in project in a most efficient manner. Let us trust in ourselves and in our organizational skills!

Chapter 4

We Want to Enforce It

Now that you have gotten to know the concept of the plug-in fields in detail, I would like to ask you: How quickly you think we can replace nuclear, coal and oil with it? I am very happy to bring you the good news now and tell you: You decide, respectively, we decide together!

With your support and the support of your friends you decide when your country begins with the implementation! You decide, by doing what is in your power, when your country will start to use the "capital of light" it has and when you will be able to supply yourself completely with economic solar energy. And you decide, when your country will become one of the most modern energy producers in the world, or one of the world's leading exporters of electricity and at the same time becomes a place with a more healthy environment and so with an even higher quality of life.

Even if your country might be predestined to implement this project first, because of the organizational and innovative capacities it has, it still needs your personal advertising for it – at this stage from the direction of the population in the direction of the politics. So, feel motivated to call or to write letters or e-mails to your members of parliament to inform them about the concept of the plug-in fields in this book. Parallel to your efforts I am doing the same. After a short search on the Internet you surely will find your delegate of your electoral district or a complete list of all current members of parliament, together with their contact details.

Also feel free to write on the personal Facebook page of the deputies. If you like, write to journalists as well, write suspenseful letters to the editor for newspapers and maybe you have even more ideas.

Let us fill our democracy with life! We thereby obtain cheap and clean energy in abundance! With the measures proposed here, we can achieve that from the date of introduction within three years! Don't you think so as well? Then support the implementation whenever and wherever you want!

On Facebook, you can follow the progress of the project at:
https://www.facebook.com/groups/photovoltaicpluginfields.international/

Also become a supporter and mobilizer on Facebook by joining the Facebook group following the above link or by searching for the group name "Photovoltaic_Plug-in_Fields_international". If you want, share the idea of this book and the link to the Facebook group with your friends and encourage them to share it as well. Or, send it to your e-mail contacts from your e-mail inbox.

Make the fast transition to a cheaper and cleaner energy possible, together with a big community of interest – for yourself and for your children! This book is the solution!

Appendices

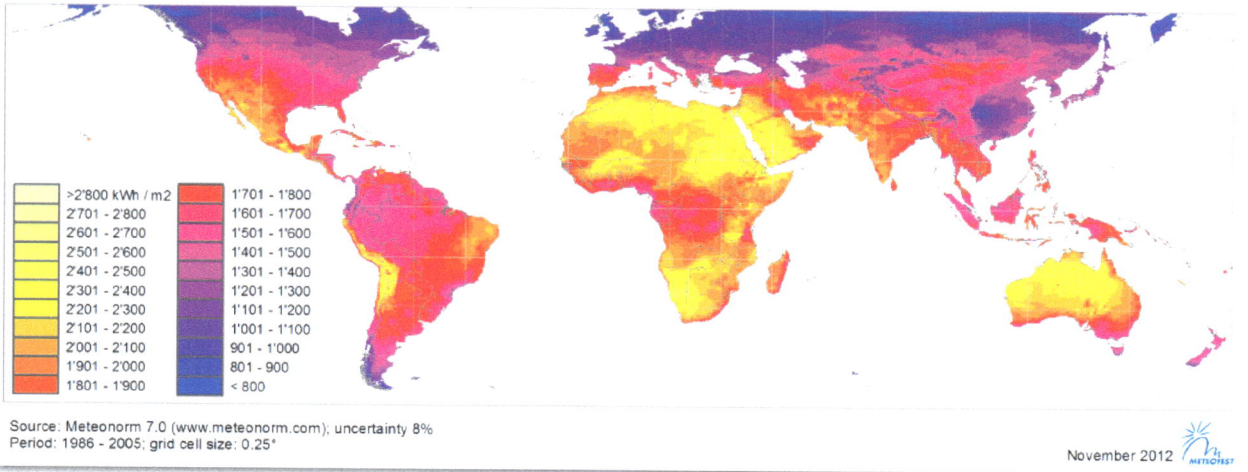

Yearly Sum of Global Horizontal Irradiation (GHI)

>2'800 kWh / m2	1'701 - 1'800
2'701 - 2'800	1'601 - 1'700
2'601 - 2'700	1'501 - 1'600
2'501 - 2'600	1'401 - 1'500
2'401 - 2'500	1'301 - 1'400
2'301 - 2'400	1'201 - 1'300
2'201 - 2'300	1'101 - 1'200
2'101 - 2'200	1'001 - 1'100
2'001 - 2'100	901 - 1'000
1'901 - 2'000	801 - 900
1'801 - 1'900	< 800

Source: Meteonorm 7.0 (www.meteonorm.com); uncertainty 8%
Period: 1986 - 2005; grid cell size: 0.25°

November 2012

(Illustration: © METEOTEST; based on www.meteonorm.com 2013)

In the above illustration you can see the solar radiation levels for your country. To find out about the approximate electricity production at the respective level of radiation you can refer to the following table:

Horizontal Solar Irradiation Level	Power Generation with Standard PV Modules Per kW peak
2,500 kWh/m²/year	2,000 kWh/year
2,000 kWh/m²/year	1,600 kWh/year
1,700 kWh/m²/year	1,400 kWh/year
1,300 kWh/m²/year	1,100 kWh/year
1,200 kWh/m²/year	1,000 kWh/year
1,100 kWh/m²/year	900 kWh/year

(Data: Kost et al. 2012, p. 11)

Bibliography

Burger, B. (2013): Fraunhofer Institute for Solar Energy Systems ISE. Electricity from solar and wind energy in 2012 (Fraunhofer-Institut für solare Energiesysteme ISE. Stromerzeugung aus Solar- und Windenergie im Jahr 2012). Freiburg, Germany. – URL: http://www.ise.fraunhofer.de/de/downloads/pdf-files/aktuelles/stromproduktion-aus-solar-und-windenergie-2012.pdf – (Download: 22.03.2013).

Energieforum-Hessen.de (2013): Infrared heaters – combined with photovoltaics. Heating with own power: infrared heaters and solar energy (Infrarotheizungen – Kombination mit Photovoltaik. Mit Eigenstrom heizen: Infrarotheizungen und Solarenergie). Frankfurt am Main, Germany. – URL: http://www.energieforum-hessen.de/infrarotheizung/co2frei-infrarot-pv-photovoltaik.html – (Download: 22.03.2013).

Federal Ministry of Economics and Technology (Bundesministerium für Wirtschaft und Technologie) (2013): Facts and figures – energy data (Zahlen und Fakten – Energiedaten). Berlin, Germany. – URL: http://www.bmwi.de/BMWi/Redaktion/Binaer/energie-daten-gesamt,property=blob,bereich=bmwi2012,sprache=de,rwb=true.xls – (Download: 14.01.2013).

Federal Statistical Office (Statistisches Bundesamt) (2012): Agricultural land is decreasing, harvests are increasing (Landwirtschaftlich genutzte Fläche rückläufig, Erntemengen legen zu). Wiesbaden, Germany. – URL: https://www.destatis.de/DE/PresseService/Presse/Pressemitteilungen/2012/10/PD12_360_412.html – (Download: 22.03.2013).

Fraunhofer-Gesellschaft (2010): Press release. Storing green electricity as natural gas (Presseinformation. Ökostrom als Erdgas speichern). München, Germany. – URL: http://www.fraunhofer.de/de/presse/presseinformationen/2010/04/strom-erdgas-speicher.html – (Download: 21.03.2013).

German Solar Industry Association (BSW – Bundesverband Solarwirtschaft e.V.) (2013): Data and information on the German solar industry (Daten und Infos zur deutschen Solarbranche). Berlin, Germany. – URL: http://www.solarwirtschaft.de/presse-mediathek/marktdaten.html – (Download: 22.03.2013).

Grupo T-Solar Global S.A. (2013): Thin-film PV Panel – TS Full SJ TS410. Madrid, Spain. – URL: http://www.tsolar.com/recursos/doc/Innovacion/Fabricacion/Modulos_TSolar_Ingles/337301033_1 9122012161919.pdf – (Download: 21.03.2013).

Houben, M. (2012): market scanner. Electric cars: How they perform and what they cost. Additional information to the program of the 10th of Sept. 2012 (markt-Scanner. Elektroautos: Was sie leisten, was sie kosten. Informationen zur Sendung vom 10.09.2012). Köln, Germany. – URL: http://www.wdr.de/tv/markt/sendungsbeitraege/2012/0910/download/Elektroautos.pdf – (Download: 22.03.2013).

Institute of Heat and Oil Technology (Institut für Wärme und Oeltechnik e. V. [IWO]) (2013): Fuel value and heat value (Brennwert und Heizwert). Hamburg, Germany. – URL: http://www.iwo.de/fachwissen/oeltechnik/brennwerttechnik/brennwert-und-heizwert/ – (Download: 22.03.2013).

Kosack, P. (2009): Report on the research project "Exemplary comparative measurement between infrared radiation heating and gas heating in old buildings" (Bericht zum Forschungprojekt „Beispielhafte Vergleichsmessung zwischen Infrarotstrahlungsheizung und Gasheizung im Altbaubereich"). Kaiserslautern, Germany. – URL: http://www-user.rhrk.uni-kl.de/~kosack/forschung/?download=ForschungsberichtIR.pdf – (Download: 22.03.2013).

Kost, C./Schlegl, T./Thomsen, J./Nold, S./Mayer, J. (2012): Study, generation costs of renewable energies. Fraunhofer Institute for Solar Energy Systems ISE (Studie Stromgestehungskosten Erneuerbare Energien. Fraunhofer-Institut für solare Energiesysteme ISE). Freiburg, Germany. – URL: http://www.ise.fraunhofer.de/de/veroeffentlichungen/veroeffentlichungen-pdf-dateien/studien-und-konzeptpapiere/studie-stromgestehungskosten-erneuerbare-energien.pdf – (Download: 03.03.2013).

Küchler, S./Litz, P. (2013): Electricity prices in Europe and the competitiveness of energy-intensive industries. Brief analysis on behalf of the Alliance 90/The Greens parliamentary group (Strompreise in Europa und Wettbewerbsfähigkeit der stromintensiven Industrie. Kurzanalyse im Auftrag der Bundestagsfraktion BÜNDNIS 90/DIE GRÜNEN). Berlin, Germany. – URL: http://www.foes.de/pdf/2013-01-Industriestrompreise-Wettbewerbsfaehigkeit.pdf – (Download: 23.03.2013).

Küchler, S./Meyer, B./Blanck, S. (2012): What power really costs. Comparison of state subsidies and society costs of conventional and renewable energies. Eco-Social Market Economy Forum (FÖS)/Greenpeace Energy eG (ed.)/German Wind Energy Association (BWE) (ed.) (Was Strom wirklich kostet. Vergleich der staatlichen Förderungen und gesamtgesellschaftlichen Kosten von konventionellen und erneuerbaren Energien. Forum Ökologisch-Soziale Marktwirtschaft e.V. [FÖS]/Greenpeace Energy eG [Hrsg.]/Bundesverband WindEnergie e.V. [BWE] [Hrsg.]). Berlin/Hamburg, Germany. – URL: http://www.foes.de/pdf/2012-08-Was_Strom_wirklich_kostet_kurz.pdf – (Download: 03.03.2013).

Municipality of Gaildorf (Stadtverwaltung Gaildorf) (2013): Natural energy storage project. Article (Projekt Naturstromspeicher. Artikel). Gaildorf, Germany. – URL: http://www.gaildorf.de/data/e-BuergerArtikel.php?id=233806 – (Download: 21.03.2013).

National Meteorological Service of Germany (Deutscher Wetterdienst) (2012): Radiation maps of the mean values (period 1981 - 2010) for Germany (Strahlungskarten der Mittelwerte (Zeitraum 1981 - 2010) für Deutschland). Offenbach, Germany. – URL: http://www.dwd.de/bvbw/appmanager/bvbw/dwdwwwDesktop?_nfpb=true&_pageLabel=dwdwww_result_page&portletMasterPortlet_i1gsbDocumentPath=Navigation%2FOeffentlichkeit%2FKlima__Umwelt%2FKlimagutachten%2FSolarenergie%2FGlobalstr__Karten__frei__node.html%3F__nnn%3Dtrue – (Download: 24.03.2013).

pvXchange GmbH (2013): Price index (Preisindex). Köln, Germany. – URL: http://www.pvxchange.com/priceindex/Default.aspx – (Download: 03.03.2013).

Renewable Energies Agency (Agentur für Erneuerbare Energien e.V.) (2012a): Electricity mix in Germany 2012 (Strommix in Deutschland 2012). Berlin, Germany. – URL: http://www.unendlich-viel-energie.de/de/bioenergie/detailansicht/article/155/strommix-in-deutschland-2012.html – (Download: 03.03.2013).

Renewable Energies Agency (Agentur für Erneuerbare Energien e.V.) (2012b): Three new energy cooperatives per week (Drei neue Energiegenossenschaften pro Woche). Berlin, Germany. – URL: http://www.unendlich-viel-energie.de/de/startseite/detailansicht/article/19/drei-neue-energiegenossenschaften-pro-woche.html – (Download: 21.03.2013).

Renewable Energies Agency (Agentur für Erneuerbare Energien e.V.) (2012c): Turnaround in energy policy reduces import dependence: renewable energies account for savings of more than 6 billion euros of energy imports (Energiewende lässt Importabhängigkeit sinken: Erneuerbare vermeiden mehr als 6 Milliarden Euro Energieimporte). Berlin, Germany. – URL: http://www.unendlich-viel-energie.de/de/detailansicht/article/4/energiewende-laesst-importabhaengigkeit-sinken-erneuerbare-vermeiden-mehr-als-6-milliarden-euro-ene.html – (Download: 22.03.2013).

Sterner, M./Jentsch, M./Trost, T./Pape, C./Gerhardt, N. (2012): Power-to-Gas: Energy storage by linking power and gas networks (Power-to-Gas: Energiespeicherung durch die Kopplung von Strom- und Gasnetzen). Regensburg, Germany. – URL: http://www.hs-regensburg.de/fileadmin/media/professoren/ei/sterner/pdf/2012_04_Sterner_Pfaffenhofen_Power-to-Gas_p.pdf – (Download: 21.03.2013).

Verivox GmbH (2013): Turn of the year: Electricity prices rose by 11 percent (Jahreswechsel: Strompreise um 11 Prozent gestiegen). Heidelberg, Germany. – URL: http://www.verivox.de/nachrichten/jahreswechsel-strompreise-um-11-prozent-gestiegen-91087.aspx – (Download: 22.03.2013).

Wirth, H. (2013): Current facts on photovoltaics in Germany. Fraunhofer Institute for Solar Energy Systems ISE (Aktuelle Fakten zur Photovoltaik in Deutschland. Fraunhofer-Institut für solare Energiesysteme ISE). Freiburg, Germany. – URL: http://www.ise.fraunhofer.de/de/veroeffentlichungen/veroeffentlichungen-pdf-dateien/studien-und-konzeptpapiere/aktuelle-fakten-zur-photovoltaik-in-deutschland.pdf – (Download: 03.03.2013).

World Nuclear Association (2012): Radioactive Wastes – Myths and Realities. London, UK. – URL: http://www.world-nuclear.org/info/inf103.html – (Download: 21.03.2013).

Illustrations

Burger, B. (2013): Fraunhofer Institute for Solar Energy Systems ISE. Electricity from solar and wind energy in 2012 (Fraunhofer-Institut für solare Energiesysteme ISE. Stromerzeugung aus Solar- und Windenergie im Jahr 2012). Freiburg, Germany. – URL: http://www.ise.fraunhofer.de/de/downloads/pdf-files/aktuelles/stromproduktion-aus-solar-und-windenergie-2012.pdf – (Download: 22.03.2013).

Energieforum-Hessen.de (2013): Infrared heaters a viable alternative? (Infrarotheizungen eine echte Alternative?). Frankfurt am Main, Germany. – URL: http://www.energieforum-hessen.de/infrarotheizung/infrarotheizungen-wissenschaftliche-studie.html – (Download: 22.03.2013).

METEOTEST (2013): Copyright maps. Bern, Switzerland. – URL: http://meteonorm.com/?id=108 – (Download: 06.05.2013).

Renewable Energies Agency (Agentur für Erneuerbare Energien e.V.) (2012d): Energy costs in private households (Energiekosten in Privathaushalten). Berlin, Germany. – URL: http://www.unendlich-viel-energie.de/de/detailansicht/article/226/energiekosten-in-privathaushalten.html – (Download: 24.03.2013).

The author

Clemens Hauser, born in 1975, is a marketing professional and an innovative developer for product and service strategies as well as for complete market scenarios. In his first book, he goes about the reduction of energy prices and the final breakthrough of renewable energies, always in the context of the feasibility and with clear proposals for action. His focus is always on the benefits for the individual as well as for the community. With this book he intends nothing less than the complete changing of the worldwide energy situation due to a new and yet obvious way of making clean energy available. Especially notable is his motivating style which is reflected in his work.

www.ingramcontent.com/pod-product-compliance
Lightning Source LLC
Chambersburg PA
CBHW052047190326
41521CB00002BA/141